R.U.R. AND THE VISION OF ARTIFICIAL LIFE

R.U.R. AND THE VISION OF ARTIFICIAL LIFE

KAREL ČAPEK

EDITED BY JITKA ČEJKOVÁ

THE MIT PRESS   CAMBRIDGE, MASSACHUSETTS   LONDON, ENGLAND

The MIT Press would like to thank the anonymous peer reviewers who provided comments on drafts of this book. The generous work of academic experts is essential for establishing the authority and quality of our publications. We acknowledge with gratitude the contributions of these otherwise uncredited readers.

This book was set in ITC Stone and Avenir by New Best-set Typesetters Ltd. Printed and bound in the United States of America.

Library of Congress Cataloging-in-Publication Data

Names: Čapek, Karel, 1890–1938, author. | Čejková, Jitka, editor. | Šimek, Štěpán S., translator.
Title: R.U.R. and the vision of artificial life / Karel Čapek ; edited by Jitka Čejková ; translated by Štěpán S. Šimek.
Other titles: R. U. R. English
Description: Cambridge, Massachusetts : The MIT Press, [2023] | Includes bibliographical references and index.
Identifiers: LCCN 2022059065 (print) | LCCN 2022059066 (ebook) | ISBN 9780262544504 (paperback) | ISBN 9780262371896 (epub) | ISBN 9780262371902 (pdf)
Subjects: LCSH: Robots—Drama. | Čapek, Karel, 1890–1938. R.U.R. | Robots in literature. | Artificial intelligence. | Artificial life. | Robotics. | LCGFT: Drama. | Literary criticism. | Essays.
Classification: LCC PG5038.C3 R213 2023 (print) | LCC PG5038.C3 (ebook) | DDC 891.8/6252—dc23/eng/20230621
LC record available at https://lccn.loc.gov/2022059065

10 9 8 7 6 5 4 3 2 1

CONTENTS

INTRODUCTION

Jitka Čejková

Nature discovered only one method of producing and arranging living matter. There is, however, another, simpler, more malleable, and quicker method, one that nature has never made use of. This other method, which also has the potential to develop life, is the one I discovered today.

Any scientist, especially one who works in the artificial life field, would love to make such a groundbreaking discovery, to be the first in the world to share these wonderful words on social media and publish the results in prestigious scientific journals. Unfortunately, another method to create life has not yet been found, and these notes were written in a laboratory book by the fictional mad scientist Rossum from Karel Čapek's play *R.U.R.*, subtitled *Rossum's Universal Robots*.

Karel Čapek (January 9, 1890–December 25, 1938) was a Czechoslovak writer, playwright, and journalist. I don't believe he ever had an ambition to be one of the "firsts" in the world, but in the end he was. With his brother Josef he invented a new word—*robot*—and he was the first to use this word for an artificial human, formed from chemically synthesized living matter. In Czech, *R.U.R.* was published in November 1920 and premiered on January 25, 1921 in the National Theatre in Prague. The play was first performed in English by the New York Theatre Guild on October 9, 1922. It was a great success—by 1923 *R.U.R.* had been translated into thirty languages. And the word "robot" remained untranslated in most of these.

Soon the word "robot" started to be used for all sorts of things, and Karel Čapek, instead of being happy at its fame, was upset and frustrated. He protested against the idea of robots in the form of electromechanical monsters that fly airplanes or destroy the world by trampling. His robots were not made of sheet metal and cogwheels; they were not a celebration of mechanical engineering! In a column in the newspaper *Lidové noviny* in June 1935 (in translation as the afterword to this book), he emphasized that when writing *R.U.R.* he was thinking instead of modern chemistry. Although it was the chemistry of the time, without concepts like DNA or RNA, his statements were quite timeless (as is all of *R.U.R.*). He laid stress on the idea that one day we will be able to produce, by artificial means, a living cell in a test tube. That we will be able to create a new kind of matter by chemical synthesis, one that behaves like living material; an organic substance, different from what living cells are made of; something like an alternative basis for life, a material substrate in which life could have evolved, had it not, from the beginning, taken the path it did. He emphasized that we do not have to suppose that all the different possibilities of creation have been exhausted on our planet. His texts were such a brilliant ode to artificial life!

However, in Čapek's lifetime there was no developed scientific field of artificial life (commonly abbreviated as ALife). This emerged several decades later. It is generally accepted that the modern field of ALife was established at a workshop held in Los Alamos in 1987 by Christopher G. Langton. The field focuses mainly on the creation of synthetic life on computers or in the laboratory, in order to study, simulate, and understand living systems. Originally the field was a conglomerate of researchers from various disciplines including computer science, physics, mathematics, biology, chemistry, and philosophy, who were exploring topics and issues far outside of their own disciplines' mainstreams, often topics of foundational and interdisciplinary character. These "renegade" scientists had problems finding colleagues, conferences, and journals in which to disseminate their research and to exchange ideas. For this rich and diverse set of people, ALife became a new home, a "big tent," unifying them especially in two main conferences (the International Conference on the Synthesis and Simulation of Living Systems, and later also the European Conference on Artificial Life, with meetings in alternating years) and one scientific journal (MIT Press's *Artificial Life*).

Today, artificial life researchers meet annually at conferences simply titled ALife. I have attended many of them, and always it thrills me to see what kinds of topics are presented and how many views on a specific problem are offered during the discussions. The common characteristic of all researchers in the ALife community is their open mind. It really is a radically interdisciplinary field that cannot be defined either as pure science or engineering.[1] It involves both, employing experimental and theoretical approaches, and the research is fundamental and mainly curiosity-driven. Practical applications come as by-products, but they are not the goal. There are so many basic questions that we are still unable to answer, such as "What is life?", "How did life originate?", "What is consciousness?" and more. Many of these questions are related not only to science but also to philosophy. And what has always fascinated me most was how many of these contemporary questions were already heard in Čapek's century-old science fiction play.

And therefore I often introduced and praised Čapek's *R.U.R.* in my talks at conferences, but also in scientific papers, in normal conversation, everywhere! Not only because I wanted to call attention to all of the fascinating open questions related to artificial life that Čapek outlined, but also to point out that the original robots were made from artificial flesh and bones, which was always surprising information for many people. Moreover, I wanted to remind the world of the Czech giant, Karel Čapek, who was nominated seven times for the Nobel Prize in literature. Of course, I also like his other works. Since childhood I have loved his books *Nine Fairy Tales* and *Dashenka, or the Life of a Puppy*. Later I read *The Makropulos Affair, The Mother, The White Disease, War with the Newts, Krakatit, The Gardener's Year*, and others. Some of these works also raise issues related to artificial life (especially *War with the Newts*), but *Rossum's Universal Robots* has become my favorite since I started my doctoral studies in the Chemical Robotics Laboratory of Professor František Štěpánek at my alma mater, the University of Chemistry and Technology Prague.

In fact, the chemical robots in the form of microparticles that we designed and investigated, and that had properties similar to living cells, were much closer to Čapek's original ideas than any other robots today ("a blob of some colloidal pulp that not even a dog would eat," as one of his characters puts it). Currently, in my laboratory I examine droplets of decanol in the environment of sodium decanoate (an organic phase almost immiscible with water in the form of a droplet located on the surface of an aqueous surfactant solution). These droplets are unique in that they somehow resemble the behavior of living organisms. For example, just as living cells or small animals can move in an oriented manner in an environment of chemical substances—in other words, they can move chemotactically (chase food or run away from poisonous substances)—so my droplets can follow the addition of salts or hydroxides in a very similar way.[2] Thanks to chemotaxis, they can even find their way out of a maze! Just as living cells change their shape and create various protrusions on their surface, so also decanol droplets are able to change their shape under certain conditions and create all kinds of tentacle-like structures.[3] We also recently discovered that amazing interactions occur in groups of many droplets, when the droplets cluster or, on the contrary, repel each other, so that their dance creations on glass slides or in Petri dishes resemble the collective behavior of animal populations like flocks.[4] Bottom line, I started to call these droplets liquid robots![5] Just as Rossum's robots were artificial human beings that only looked like humans and could imitate only certain characteristics and behaviors of humans, so liquid robots, as artificial cells, only partially imitate the behavior of their living counterparts.

As I mentioned above, *R.U.R.* was published in 1920. As the year 2020 approached, I felt that we must mark the centenary of this timeless work and celebrate the word "robot" in some way. Some of my ideas were never realized, and some were only partially realized due to the COVID pandemic (we organized the ALife 2021 conference as an online-only event, rather than hosting it in Prague as we wanted). The project I really took seriously was to prepare a book on the occasion of the 100th anniversary of the word "robot." This book would contain Čapek's original play along with present-day views on this century-old story.

And thus the book *Robot 100: Sto rozumů* was released by our University of Chemistry and Technology Prague in November 2020, exactly 100 years after Čapek's *Rossum's Universal Robots*. It contained the contributions of 100 people, mostly scientists, but also writers, journalists, radio and television presenters, musicians, athletes, and artists. *R.U.R.* is a timeless work, in which we can find many topics that scientists deal with even today, whether the synthesis of artificial cells, tissues, and organs, issues of evolution and reproduction, or the ability to imitate the behavior of human beings and show at least signs of intelligence or consciousness. *R.U.R.* also outlines social problems related to globalization, the distribution of power and wealth, religion, and the position of women in society. Every contributor could find an example of how *R.U.R.* raises some still unanswered questions of their field.

The feedback of readers and the positive reviews encouraged me to publish an English edition. I was pleased that the MIT Press was interested in publishing my book. The key problem, as I had already seen in feedback from contributors, was the English translations of Čapek's play. Probably the most widely used English translation of *R.U.R.* is the very first one from 1923 by Paul Selver. However, it is not a very successful translation, and Čapek himself was not satisfied with it. Not only did Selver leave out some passages, but he even completely canceled the character of the robot Damon. Also, while Čapek's original consists of a prologue and three acts, in translations we often encounter three acts and a final epilogue. Foreign authors often wrote about Rossum junior as a son because he is referred to as "young Rossum," but in the Czech original it is clearly stated that he was the nephew of the older Rossum. Another difference was Domin's request for Helena Glory's hand—while in the Czech original he only places both hands on her shoulders, in Selver's translation he even kisses her. Inconsistencies in the translations were mostly easily solvable trifles, but sometimes they complicated the content of the entire essay.

It was clear that if we published *Robot 100* in English, it would require a completely new translation. I am happy that Professor Štěpán Šimek agreed to translate the first edition of Čapek's *Rossum's Universal Robots* for this book. I was so happy when I obtained his comments to the translation: "This is a straight translation in terms of the original. By that I mean, that unlike some other translations that I'm familiar with, I have not made any cuts, that I have translated every word and line, and that I haven't changed anything from the original. While the play has some obvious dramaturgical flaws, I have not tried to correct those in the translation. I believe that cuts, rearrangements, dramaturgical clarifications, and stuff like that are the job of the potential director and/or dramaturg, not the translator, unless the translator is asked to create an adaptation of the original. In other words, this is a Translation, not an Adaptation."

And I was amazed when I read the new translation. It is excellent and it fulfills my expectations. Thanks to Štěpán Šimek's work, this book offers English-speaking readers a truly faithful translation of Čapek's *R.U.R.*

It was obvious that if we published book *Robot 100* in English in 2023, it would require a new title, because there is no centenary this year. We also decided not to include the contributions of all one hundred of the contributing authors in our print edition, but to select mainly those related to artificial life, and to publish the rest online. Finally, we have chosen the title *R.U.R. and the Vision of Artificial Life* for the book that you have in your hands right now. This title perfectly reflects what the readers will find within—the century-old play *R.U.R* in a completely new and modern translation by Štěpán Šimek, and twenty essays on how Čapek's brilliant play has the prescient power to illustrate current directions and issues in artificial life research and beyond.

R.U.R. (ROSSUM'S UNIVERSAL ROBOTS)

Karel Čapek
translated by Štěpán S. Šimek

CHARACTERS

HARRY DOMIN   Chief Executive Officer, Rossum's Universal Robots Company (R.U.R.). In his mid to late thirties, tall, clean-shaven.

FABRY   Chief Technology Officer (R.U.R.). Earnest-looking and delicate, pale, also clean-shaven.

DR. GALL   Director of the Office of Physiological Research and Development (R.U.R.). Small and lively, tanned with a black moustache.

DR. HALLEMEIER   Director of the Office of Psychology and Education of Robots. A large and loud redhead with a crew cut and a moustache.

BUSMAN   Chief Financial Officer (R.U.R). Fat, bald, and nearsighted.

ALQUIST   Director of Physical Plant and New Construction. Older than the others, shabbily dressed, with long graying hair and full beard.

HELENA GLORY   The Daughter of President Glory, later, Harry Domin's wife. Young and stylishly dressed.

NANA   Helena's old nurse

MARIUS   A male Robot

SULLA   A female Robot

RADIUS   A male Robot

DAMON   A male Robot

PRIMUS   A male Robot

HELENA   A female Robot

OTHER ROBOTS

The ROBOTS are dressed just like real humans in the Prologue. Their movements and inflections are choppy, their faces are expressionless, and they stare emotionlessly. In the rest of the play, they wear linen shirts tied by a belt above their hips, and a large brass number attached to their chests.

PROLOGUE

The executive office of Rossum's Universal Robots factory. At the right is the main entrance. Large windows in the back through which we see a seemingly never-ending row of industrial buildings. Other executive offices on the left.

DOMIN is sitting in an office chair at a large executive-style desk. On the desk are a desk lamp, a telephone, paperweights, a box of letters, etc.; on the wall on the left hang large naval and railroad maps, a large calendar, and a clock showing just before noon; pinned on the wall on the right are various print ads: "The Cheapest Labor—Rossum's Robots," "Tropical Robots—the latest invention. $150 apiece," "Buy your own personal Robot," "Want to produce cheaper? Buy Rossum's Robots." There are other maps, ship schedules, a table with the latest currency exchange rates, etc. In contrast to the eclectic wall decorations are an expensive Turkish rug on a parquet floor; on the right a round coffee table, a small sofa, leather armchairs, and a bookshelf filled not with books, but rather with bottles of wine and spirits. On the left is a safe. A young woman, SULLA, is typing on a typewriter next to DOMIN's desk.

DOMIN [*dictates*]: ". . . that we don't assume responsibility for goods damaged in transit. We called your captain's attention to the fact that the vessel was unsuitable for the transportation of Robots, and therefore we are not liable for the loss of the cargo. Signed—on behalf of Rossum's Universal Robots—" Done?

SULLA: Yes, sir.

DOMIN: Next sheet. "E. B. Huysum Agency, New York. Date. This is to confirm your order of five thousand Robots. Since you are sending your own vessel, we request that you load the ship with a cargo of coal briquettes for R.U.R. as an advanced payment for your order. Signed . . ." Done?

SULLA [*finishing typing*]: Yes, sir.

DOMIN: Next. "To Friedrichswerke, Hamburg. Date. This is to confirm your order of fifteen thousand Robots . . ." [*The phone rings. DOMIN picks it up.*] Central Office, Domin speaking . . . Yes . . . Certainly . . . Yes, of course, just as usual . . . Absolutely, let them know . . . Very well. [*He hangs up.*] Where were we?

SULLA: This is to confirm your order of fifteen thousand R.

DOMIN [*lost in thought*]: Fifteen thousand R . . . Fifteen thousand R.

[MARIUS *enters.*]

MARIUS: Sir, there is a lady waiting . . .

DOMIN: Who is it?

MARIUS: I do not know, sir. [*He hands him a calling card.*]

DOMIN [*reading the card*]: President Glory . . . Let her in.

MARIUS [*opening the door*]: This way, madam.

[HELENA GLORY *enters.* MARIUS *leaves.*]

DOMIN [*getting up*]: Come in please.

HELENA: General Director Domin, I presume?

DOMIN: At your service.

HELENA: I've come with . . .

DOMIN: . . . President Glory's calling card in your hand. No need to explain.

HELENA: President Glory is my father. I am Helena Glory.

DOMIN: Miss Glory, it's an extraordinary honor to . . . to . . .

HELENA: . . . to *not* show you the door . . . ?

DOMIN: . . . to welcome the daughter of a great president. Please, sit down. Sulla, you may leave.

[SULLA *leaves.*]

DOMIN [*sits down*]: Well, what can I do for you, Miss Glory?

HELENA: I'd like to . . .

DOMIN: . . . see our industrial production of humans. Just like every other visitor. Of course, with pleasure.

HELENA: I thought it was forbidden . . .

DOMIN: . . . to tour the factory. Yes, absolutely. But then again, everybody comes here armed with somebody's calling card or other, Miss Glory.

HELENA: And you show it to everyone?

DOMIN: Only some of it. The manufacture of artificial humans, miss, is an industrial secret.

HELENA: You can't imagine how . . .

DOMIN: . . . incredibly interesting this is for me. It's all Old Europe is talking about.

HELENA: Why don't you let me finish?

DOMIN: Please forgive me. Did you want to say something different by any chance?

HELENA: I just wanted to ask . . .

DOMIN: . . . if I could make an exception and let you see our factory. But of course, Miss Glory.

HELENA: How did you know what I wanted to ask?

DOMIN: Everyone asks that. [*Getting up.*] As a special honor, Miss Glory, we'll show you more than to the others, if . . .

HELENA: Thank you.

DOMIN: . . . if you promise never to divulge, not to anybody, even the smallest . . .

HELENA [*stands up and takes his hand*]: You have my word.

DOMIN: Thank you. Would you mind lifting your veil?

HELENA: Ah, of course, you need to see . . . I'm sorry.

DOMIN: I beg your pardon?

HELENA: You may want to let go of my hand.

DOMIN: [*lets go of her hand*]: Please forgive me.

HELENA [*lifting her veil*]: I see. You want to make sure I'm not a spy. How suspicious you are!

DOMIN [*enchanted by what he sees*]: Hmm . . . Of course . . . we . . . well, so be it . . .

HELENA: Don't you trust me?

DOMIN: Exceedingly, Miss Hele . . . pardon me, Miss Glory. Exceedingly pleased, indeed . . . Did you have a good crossing?

HELENA: Yes. Why . . .

DOMIN: Because . . . what I mean to say is . . . that you're still very young.

HELENA: Well, are we going to see the factory now?

DOMIN: Absolutely. Twenty-two, I guess. Am I correct?

HELENA: Twenty-two what?

DOMIN: Years old.

HELENA: Twenty-one. Why do you want to know?

DOMIN: Because . . . since . . . [*with excitement*] You'll stay for a while, won't you?

HELENA: Depends on what you show me in the factory.

DOMIN: That damned factory! But of course, Miss Glory, you'll see everything. Are you interested in the history of the invention?

HELENA: Oh yes, please. [*She sits down.*]

DOMIN: Well then. [*He sits on the desk, and rapturously watching* HELENA, *quickly reels off.*] It was in the year 1920, when the old Rossum, a great physiologist, but then still a young scholar, departed for this remote island in order to study marine organisms, period. As part of his research, he undertook to replicate a living material, called protoplasm, by the process of chemical synthesis until, by chance, he discovered a matter that behaved exactly like living material, even though its chemical composition was markedly different, and that was in the year 1932, exactly four hundred and forty years after the discovery of America, oof.

HELENA: Did you memorize that?

DOMIN: I did. Physiology, Miss Glory, is not exactly my strength. Should I go on?

HELENA: I don't mind.

DOMIN [*triumphantly*]: And it was then, miss, when old Rossum wrote the following note in the margin of his chemical equations notebook: "Nature discovered only one method of arranging living matter. There is, however, another, simpler, more malleable, and quicker method, one that nature has never made use of. This other method, which also has the potential to develop life, is the one I discovered today." You have to imagine, miss, that he wrote those great words while contemplating a blob of some colloidal pulp that not even a dog would eat. Imagine him sitting over a test tube and already seeing the whole tree of life growing out of it, and every animal, starting with some rotifer and ending with a human being, emerging from it. Ending with a human being! A human made of a different material than us. That, Miss Glory, that was a tremendous moment.

HELENA: And then?

DOMIN: Then? Then the question was how to get that life out of the tube, how to accelerate the development and create some of the organs, bones, nerves, and what have you, and to make all kinds of substances, such as catalysts, enzymes, hormones, and so on . . . in short . . . do you understand it?

HELENA: N . . . no, I don't. I mean, maybe some of it.

DOMIN: Well, I personally don't understand any of it. But you know, by using all of those little concoctions of his, he could have made anything he wanted. He could have made, I don't know, a jellyfish with the brain of a Socrates, or a fifty-yard-long earthworm. But since he didn't have a funny bone in his body, he obsessed about making regular vertebrates, or maybe even a human. That artificial living material of his had an insane will to live; it tolerated everything—he could stitch it and mix it together in any way he wanted. Which is something that you could never do with a naturally occurring protein. So, he went to work.

HELENA: To do what?

DOMIN: To imitate nature. First, he tried to make an artificial dog. It cost him years of work, and all that came out of it was a sort of stunted calf-like creature that croaked only a few days later. I'll show it to you in our museum. And then, finally, old Rossum set out to make a human.

[Pause]

HELENA: And I'm not to divulge this to anybody?

DOMIN: To nobody in the world.

HELENA: Shame that it's already in all our textbooks.

DOMIN: Shame indeed. [He hops off the table, and sits down next to HELENA.] But you know what is not in the textbooks? [He taps the side of his head.] That old Rossum was stark raving mad. But really, Miss Glory, this one you must keep to yourself. That old maniac really wanted to make people.

HELENA: But that's exactly what you're doing!

DOMIN: In a sense, yes, Miss Helena. But old Rossum took it literally. You could say that he wanted to scientifically dethrone God. He was a horrible old materialist, and that's why he did what he did. He wanted nothing less than to prove that there's no need for God, and that there never was. That's why he got it into his head to make an exact replica of a human being, to make it just like us. Do you know anything about anatomy?

HELENA: A bit—not much.

DOMIN: Neither do I. But you have to imagine that he was fixated on making everything to the last gland just like in a human body. Appendix, tonsils, belly button, all that useless stuff. Even the . . . ehm . . . the sexual organs.

HELENA: But those . . . those really are . . .

DOMIN: . . . not useless, I know. But if people are to be artificially manufactured, then you don't . . . ehm . . . one doesn't need to . . .

HELENA: I get it.

DOMIN: I'll show you that mangled mess that took him ten years to produce in the museum later on. It was supposed to be a man, but it lived for barely three days. Old Rossum had zero taste, and it looked horrible. But on the inside, it had everything you find in a human body. The work of a real stickler. And just around that time, the old man's nephew, engineer Rossum, came here to join him. A real genius, Miss Glory. The moment he saw what the old fool was trying to concoct, he told him: "Spending ten years making a human is absurd. If you can't make it quicker than nature, then you may just as well close up the shop." And then he delved into the anatomy himself.

HELENA: The textbooks tell a different story.

DOMIN [*getting up*]: Your textbooks are full of paid ads, and just between you and me, they're all wrong. For example, I'm sure that they taught you that the Robots were old Rossum's invention. But in reality, while the old man would have made a perfect university researcher, he didn't have a clue about industrial manufacturing. He thought that he'd make real people, like, I don't know, some new race of desert dwellers, or intellectuals, or, for all I care, idiots, you know what I mean? But it was only young Rossum who realized that you can transform all that mess into the manufacturing of living, intelligent labor machines. All that stuff in your textbooks about the "collaboration" between the two great Rossums is pure fiction. They were at each other's throats all the time. The old atheist had absolutely no patience for industry, and so finally, the young Rossum basically locked him up in some lab in the basement to let him fiddle with his monumental abortions, while he himself took over the production like a *real* engineer. Old Rossum literally cursed him, and before he died, he managed to patch together two more physiological monstrosities, until finally they found him dead in his lab. And that's the whole story.

HELENA: And what about the young one?

DOMIN: Young Rossum, miss, he was the new age. The age of manufacturing surpassing the age of knowledge. Just having briefly glanced at human anatomy, he quickly realized that it was unnecessarily complicated, and that any decent engineer could make it far simpler. So, he started to play around with it, he experimented with what could be left out, what could be simplified, and so on—in one word, Miss Glory, am I boring you?

HELENA: No, on the contrary, it's incredibly fascinating.

DOMIN: Well then, so, young Rossum is thinking: A human being is something that, let's say, experiences joy, plays the violin, feels like taking a walk, and in general needs to do tons of things, which . . . which are essentially completely useless.

HELENA: Oh, come on!

DOMIN: Wait. Let me explain. Which are useless when, for example, they have to weave a cloth or perform addition. I don't think that as far as you are concerned . . . Do you play the violin?

HELENA: No, I don't.

DOMIN: Too bad. But a working machine doesn't need to play the violin, it doesn't need to be happy, it doesn't need to do plenty of other things either. As a matter of fact, it shouldn't do them. A diesel engine shouldn't have tassels and artistic ornamentation, Miss Glory. And manufacturing artificial workers is just like manufacturing diesel engines. The production ought to be as streamlined as possible, and the product ought to be perfect. What do you think makes a perfect worker?

HELENA: The perfect worker? Well, I guess one, who . . . who is honest . . . and loyal.

DOMIN: Wrong. One who is the cheapest. One that has the least needs. Young Rossum invented a worker with the smallest number of needs. He had to simplify him. He tossed out everything not directly related to the task at hand, and by doing that, he essentially kicked out a human being, and created a Robot. My dear Miss Glory, Robots are not people. They are mechanically far superior to us, they have an astonishing capacity to reason intelligently, but they don't have a soul. Have you ever seen the inside of a Robot?

HELENA: No.

DOMIN: It's very clean and very simple. A wonderful piece of work, really. Like a first aid kit—only a few items, but flawlessly organized and completely functional. Ah, Miss Glory, a creation designed by an engineer will always be technically more refined than a creation of nature.

HELENA: They say that humans are the creation of God.

DOMIN: All the worse for them. God had no idea about modern technology. Did you know that young Rossum actually did want to play God?

HELENA: What did he do?

DOMIN: He started experimenting with Über-robots. Gigantic workers. He tried to make them up to twelve feet tall, but you can't imagine how those behemoths kept snapping.

HELENA: Snapping?

DOMIN: Yes, snapping. Just out of nowhere a leg would crack, or something. It seems like our planet is a bit too small for giants. So now we stick to naturally sized and more or less human-looking Robots.

HELENA: I saw the first Robots back home. The town bought them . . . I mean it hired them to . . .

DOMIN: Bought them, my dear Lady. Robots are bought.

HELENA: . . . acquired them as street cleaners. I saw them sweeping. They are so strange, so quiet . . .

DOMIN: Did you see my secretary?

HELENA: I didn't notice.

DOMIN [ringing]: You see, Rossum's Universal Robots factory doesn't just produce a standardized line of Robots. Some of our Robots are extremely refined, others are simpler and rougher. The better ones may live up to twenty years.

HELENA: And then they die?

DOMIN: Yes, they wear out.

[SULLA *enters.*]

DOMIN: Sulla, introduce yourself to Miss Glory.

HELENA [*stands up and offers* SULLA *her hand*]: Nice to meet you. You must be very sad to be so far away from home, aren't you?

SULLA: I do not know, Miss Glory. Please take a seat.

HELENA [*sits down*]: Where are you from, miss?

SULLA: I am from here, from the factory.

HELENA: You were born here?

SULLA: Yes, I was made here.

HELENA [*jumps up*]: What??!!

DOMIN [*laughing*]: Sulla is not a human, miss. Sulla is a Robot.

HELENA: I'm so sorry . . .

DOMIN [*puts his hand on* SULLA'*s shoulder*]: Sulla's not upset. Just look at her complexion, Miss Glory. Touch her face.

HELENA: Oh, no, no!

DOMIN: You couldn't tell the difference between the real stuff and this material. Look, she's even got the characteristic peach fuzz of a blonde. OK, the eyes are a bit . . . But the hair, on the other hand! Turn around, Sulla!

HELENA: Stop it!

DOMIN: Talk with our guest, Sulla. She is a very special visitor.

SULLA: Please sit down, miss. [*They both sit down.*] Did you have a good crossing?

HELENA: Yes . . . of . . . course.

SULLA: Do not return on the *Amelia* tomorrow, Miss Glory. The atmospheric pressure is falling significantly, down to seven-o-five. Wait for the *Pennsylvania*, which is a very good and a very powerful vessel.

DOMIN: Specify.

SULLA: Twenty knots per hour. Net register tonnage: two hundred thousand. One of the newest vessels, Miss Glory.

HELENA [*stutters*]: Thank . . . you.

SULLA: Crew of eighty, Captain Harpy, eight furnaces . . .

DOMIN [*laughing*]: That's enough Sulla. Now show us your French.

HELENA: You know French?

SULLA: I know four languages. I write: Dear sir! Sehr Geehrter Herr! Monsieur! Ctěný pane!

HELENA [*jumps up*]: That's outrageous! You're a charlatan! Sulla isn't a Robot, Sulla is a young woman, just like me! Sulla, this is reprehensible— why are you going along with this farce?

SULLA: I am a Robot.

HELENA: No, no, you're lying! Oh Sulla, forgive me, I know . . . they forced you into acting as a Robot for their advertisements. Sulla, you're a young woman, just like me, right? Say it!

DOMIN: I'm sorry, Miss Glory, but Sulla is a Robot.

HELENA: You're a liar!

DOMIN [*stands up abruptly*]: Oh really? [*Rings.*] I'm sorry, miss; in that case I have to convince you.

[MARIUS *enters.*]

DOMIN: Marius, take Sulla to the dissection room, and have her opened up. Quick!

HELENA: To where?

DOMIN: To the dissection room. When they cut her open, I'll take you to see her.

HELENA: No, you won't!

DOMIN: I'm sorry, but you called me a liar.

HELENA: Are you going to have her killed?

DOMIN: Machines don't get killed.

HELENA [*holding* SULLA *close*]: Don't worry, Sulla, I won't let them hurt you! Tell me, darling, are they all so inhumane to you? You mustn't allow that; please listen to me! You mustn't, Sulla!

SULLA: I am a Robot.

HELENA: That doesn't matter. Robots are good people, just like we are. Sulla, you'd really let them cut you open?

SULLA: Yes.

HELENA: Oh, you're not afraid to die?

SULLA: I do not know, Miss Glory.

HELENA: Do you know what would happen to you?

SULLA: I would stop moving.

HELENA: This is horrible!

DOMIN: Marius, tell Miss Glory what you are.

MARIUS: Robot Marius.

DOMIN: Would you take Sulla to the dissection room?

MARIUS: Yes.

DOMIN: Would you feel sorry for her?

MARIUS: I do not know.

DOMIN: What would happen to her?

MARIUS: She would stop moving. They would dump her into the crusher.

DOMIN: That means she'd die. Marius, are you afraid of dying?

MARIUS: No.

DOMIN: Well, you see, Miss Glory. Robots don't cling to life. They don't have anything to cling with because they don't "enjoy" life. They're less alive than your croquet lawn.

HELENA: Oh, stop it! At least send them away!

DOMIN: Marius, Sulla, you can leave now.

  [SULLA and MARIUS exit.]

HELENA: They are horrifying. It's disgusting what you're doing here!

DOMIN: Why disgusting?

HELENA: I don't know. Why . . . why did you call her Sulla?

DOMIN: You think it's an ugly name?

HELENA: It's a man's name. Sulla was a Roman general.

DOMIN: Oh, we thought that Marius and Sulla were lovers.

HELENA: No, Marius and Sulla were Roman generals and they actually fought against each other in the year . . . the year . . . I don't remember.

DOMIN: Come over to the window. What do you see?

HELENA: Bricklayers.

DOMIN: Those are Robots. All our workers are Robots. And can you see what's down there?

HELENA: Some office or something.

DOMIN: The business office. And . . .

HELENA: . . . it's full of office workers.

DOMIN: Robots. All of our office staff are Robots. When you see the factory . . .

[*The factory whistles go off.*]

DOMIN: Noon break. Robots don't know when to stop working. I'll show you the vats in the afternoon.

HELENA: What vats?

DOMIN [*businesslike*]: The kneading vats for the batter. Each vat contains material for up to a thousand Robots. Then there are the containers for the liver, the brains, and so on. Later on, you'll see the bone factory, and then I'll take you to the spinning mill.

HELENA: Spinning mill?

DOMIN: Yes, a nerve spinning mill, the artery and capillary spinning mill, a mill where we spin hundreds of miles of intestines, and so on. Then there is the assembly line where the whole enchilada is put together, you know, like in a car factory. Each worker installs just one part, and then the belt moves it to another one, and another, and so on, ad infinitum. That part is the most interesting one to watch. And finally comes the drying room and the storage room where the new products are put to work.

HELENA: Jesus, they're put to work right away?

DOMIN: Well. Work . . . Basically it's like when you buy new furniture. They are getting used to being. They somehow congeal on the inside or something like that. They even grow some new things inside. We need to leave a bit of room for natural development, you know. And in the meantime, the products are cultivated.

HELENA: What's that?

DOMIN: It's like going to "school." They learn how to speak, how to write and count. The thing is, they have amazing memory. You could read a twenty-volume encyclopedia to them, and they'd repeat it word for word. But they can never come up with anything new or original. They'd make great college professors, now that I think about it. And then they are sorted and shipped, about fifteen thousand pieces a day, excluding the relatively stable percentage of defective specimens, which are tossed into the crusher . . . and so on and so on . . . satisfied?

HELENA: You seem to be angry with me.

DOMIN: Of course I'm not! I would just like to . . . to maybe talk about something other than Robots. There are only a few of us here among the hundreds of thousands of Robots, and not a single woman. All we ever talk about is production, the whole day, every day, day in, day out . . . It's like the isle of the damned, Miss Glory.

HELENA: I'm sorry that I said, that . . . that . . . that you were a liar . . .

[*There is a knock on the door.*]

DOMIN: Come in, boys.

[FABRY, DR. GALL, DR. HALLEMEIER, *and* ALQUIST *enter from the left.*]

GALL: Excuse us, I hope we're not interrupting anything.

DOMIN: Come in, come in! [*Introducing everyone.*] Miss Glory, this is Alquist, Fabry, Gall, Hallemeier. The daughter of President Glory.

HELENA [*a bit at a loss*]: Good afternoon.

FABRY: We had no idea . . .

GALL: It's a tremendous pleasure . . .

ALQUIST: Welcome, Miss Glory.

[BUSMAN *bursts in from the right excitedly.*]

BUSMAN: Hello!? What's going on here?

DOMIN: Calm down, Busman. [*Introduction.*] This is our very own Mr. Busman, miss. The daughter of President Glory.

HELENA: Pleased to meet you.

BUSMAN: Oh my, oh my, what an honor! We need to notify the press that you've graced us with your presence here, if that's OK with you, Miss Glory . . .

HELENA: Oh no, no, please don't!

DOMIN: Please sit down, miss.

FABRY, BUSMAN, GALL [*speaking over each other and bumping into each other as they're pulling up chairs to sit closer to* HELENA]: After you . . . Be so kind . . . I'm so sorry . . .

ALQUIST: How was your crossing, Miss Glory?

GALL: Will you stay with us for a while?

FABRY: How do you like the factory, Miss Glory?

HALLEMEIER: Did you come on the *Amelia*?

DOMIN: Quiet! Let Miss Glory speak.

HELENA [*to* DOMIN]: What should I talk to them about?

DOMIN [*puzzled*]: Anything you want to.

HELENA: Should I . . . may I speak completely openly?

DOMIN: But of course.

HELENA [*hesitates, but then with a desperate determination*]: Tell me, aren't you ever embarrassed by the way they treat you?

FABRY: Who, if I may ask?

HELENA: The people.

[*The men look at each other bewildered and at a loss as to what to say.*]

ALQUIST: How they treat *us*?

GALL: Why would you think so?

HALLEMEIER: Well, I'll be damned!

BUSMAN: God have mercy, Miss Glory!

HELENA: But don't you feel that your existence could be much better?

GALL: It depends, Miss Glory. What do you mean?

HELENA: I think that . . . [*She explodes.*] . . . this is monstrous! This is outrageous! [*Stands up.*] All of Europe is talking about what they are doing to you here! That's why I came here, to see it for myself, and it's a thousand times worse than people think! How can you put up with it?

ALQUIST: Put up with what?

HELENA: With your situation. For crying out loud, aren't you people like the rest of us, like all of Europe, like the whole world!? The way you live here is scandalous and undignified!

BUSMAN: My goodness, Miss Glory!

FABRY: No, really boys, she's not wrong. We live here like savages.

HELENA: Worse than savages! May I, oh, may I call you brothers?

BUSMAN: By God, Miss Glory, of course you can!

HELENA: Brothers, I didn't come here as the President's daughter. I came here as a representative of the League of Humanity. Brothers, the League of Humanity has already more than two hundred thousand members. Two hundred thousand people are standing with you and are ready to help you.

BUSMAN: Two hundred thousand people. Wow! That's not too shabby, that's actually quite wonderful.

FABRY: That's what I keep saying, there's nothing better than good old Europe. You see, she hasn't forgotten us, she's offering us help.

GALL: What help? A theater?

HALLEMEIER: An orchestra?

HELENA: More than that.

ALQUIST: You, yourself?

HELENA: Oh no, I'm not important! But I'll stay as long as I'm needed.

BUSMAN: God in heaven, how delightful!

ALQUIST [to DOMIN]: Harry, I'm going to set up the best room in the house for the miss.

DOMIN: Wait. I'm afraid that . . . that Miss Glory isn't done speaking yet.

HELENA: No, I'm not done. Unless you shut me up by force.

GALL: Don't even think about it, Harry!

HELENA: Thank you. I knew you would stand by me.

DOMIN: I beg your pardon, Miss Glory. But are you sure you're talking to Robots?

HELENA [*stops dead*]: Who else would I talk to?

DOMIN: Well, I'm sorry Miss Glory. These gentlemen are actually people just like you. Just like all of Europe.

HELENA [*to the others*]: You aren't Robots?

BUSMAN [*splitting his sides*]: God protect us!

HALLEMEIER: Ugh, Robots!

GALL [*laughing*]: Thank you very much, how very kind of you!

HELENA: But . . . that's impossible!

FABRY: Upon my honor, miss, we're not Robots.

HELENA [*to* DOMIN]: So why did you tell me that all your office workers are Robots?

DOMIN: Office workers, yes. But not the executives. Miss Glory, allow me to introduce Fabry, an engineer and Chief Technology Officer of Rossum's Universal Robots; Doctor Gall, the director of the Office of Physiological Research and Development; Doctor Hallemeier, the director of the Office of Psychology and Education of Robots; Counselor Busman, Chief Financial Officer; and General Contractor Alquist, the Director of the Physical Plant and New Construction of Rossum's Universal Robots.

HELENA: Please forgive me, gentlemen, that I . . . that I . . . This is awful, what have I done!?

ALQUIST: You've done nothing wrong, Miss Glory, please sit down.

HELENA [*sitting down*]: I'm such a silly girl. And now . . . now you'll send me packing on the first boat.

GALL: Absolutely not, miss. Why on earth would we do something like that?

HELENA: Because now you know . . . because . . . because you know that I'd stir up trouble with the Robots.

DOMIN: Dear Miss Glory, we've already had hundreds of saviors and prophets visiting us here. Boatloads of missionaries, anarchists, Salvation Army soldiers, and what have you. You wouldn't believe how many religions and crazies there are in the world.

HELENA: And you let them talk to the Robots?

DOMIN: What's the harm? So far, every one of them gave up. The Robots remember everything they tell them, but that's all, nothing beyond that.

They don't even laugh at the nonsense those people are telling them. It's almost unbelievable. But if that's something you'd like to try, dear miss, I'll take you to the warehouse. You'll have a captive audience; there are about three hundred thousand Robots in there.

BUSMAN: Three hundred forty-seven thousand.

DOMIN: OK, then. You can lecture them on anything you want. You can read them the Bible, or teach them trigonometry, whatever you please. You can even preach to them about human rights.

HELENA: Oh, I think that . . . maybe if one were to show them a little love . . .

FABRY: Impossible, Miss Glory. Nothing is more unlike a human than a Robot.

HELENA: If that's so, why do you even make them?

BUSMAN: Haha, that's a good one. Why does one make Robots!

FABRY: To work, Miss Glory. A single Robot replaces two and a half workers. The human machine, Miss Glory, is remarkably wasteful. Eventually, it had to be discarded, once and for all.

BUSMAN: It was too expensive.

FABRY: It wasn't efficient enough. It just couldn't keep up with modern technology. Moreover . . . moreover . . . It is a great technological leap to . . . forgive me.

HELENA: What?

FABRY: Please forgive me. It is a great technological leap to be able to conceive and give birth by a machine. It's faster and more convenient. Any increase in speed is progress, my dear, but nature doesn't have a clue about modern labor efficiencies. From a technical point of view, the whole idea of childhood, for example, is completely absurd. It's a waste of time, and an untenable drain on resources, Miss Glory. And third . . .

HELENA: Oh, stop it already!

FABRY: Of course, I'm sorry. By the way, what does that league . . . that league . . . that League of Humanity of yours actually want to accomplish?

HELENA: We're especially concerned . . . especially concerned with the protection of Robots, and we want to ensure that they are well taken care of.

FABRY: That's not a bad goal. Machines ought to be well taken care of. I wholeheartedly approve of that. I hate it when things get broken carelessly. In that case, Miss Glory, please sign us up as regular contributing members of your League!

HELENA: No, you misunderstand me. Our goal is . . . especially to . . . our goal is to free the Robots.

HALLEMEIER: Come again?

HELENA: One ought to treat them . . . treat them like people.

HALLEMEIER: Oh. Should they be voting in elections? Should they be drinking beer? Should they be ordering us around?

HELENA: Why shouldn't they be allowed to vote?

HALLEMEIER: Should they even get paid a wage?

HELENA: Of course they should!

HALLEMEIER: That's interesting. And what would they do with the money, if I may ask?

HELENA: They'd buy . . . anything they need . . . something that'd please them.

HALLEMEIER: That's all very nice, Miss Glory, but Robots don't get pleased. I mean, really, what would they buy? You could feed them pineapples, or hay, or whatever; they don't care, they don't have a sense of taste. They're not interested in anything, Miss Glory. Hell, nobody has yet seen a Robot smile, and nobody ever will.

HELENA: Why . . . why . . . why don't you make them happy?

HALLEMEIER: That's not the way it works, Miss Glory. They are just Robots.

HELENA: Oh, but they have all this intelligence.

HALLEMEIER: That damned intelligence, yes, they do have that, miss, but nothing more. They are without free will, without passion, without history, without a soul.

HELENA: Without love and defiance?

HALLEMEIER: That goes without saying. Robots don't love anything, not even themselves. And defiance? I don't know; there are some rare instances, every so often . . .

HELENA: What?

HALLEMEIER: It's nothing, really. Sometimes they have some sort of a fit. Something like epilepsy, you know? We call it a Robotic spasm. All of a sudden, a Robot throws whatever's in its hand on the floor, and just stands there gnashing its teeth. We throw those into the crusher, of course. Probably some organic malfunction.

DOMIN: A manufacturing defect. It needs to be fixed.

HELENA: No, no, it's the soul!

FABRY: You think that gnashing one's teeth is the beginning of a soul?

HELENA: I don't know. Maybe it's some sort of a rebellion. Maybe it's even a sign that they're wrestling with something. Oh, if only you could find a way to light that spark in them!

DOMIN: It will be fixed, sooner or later, Miss Glory. Right now, Doctor Gall is experimenting with . . .

GALL: . . . not with that issue, Harry; right now, I'm working on pain receptors.

HELENA: Pain receptors?

GALL: Yes. The Robots are insensitive to pain. Young Rossum oversimplified the nervous system, and that turned out to be a problem. We need to introduce suffering.

HELENA: Why . . . why . . . If you wouldn't give them a soul, why would you want to give them pain?

GALL: For industrial reasons, Miss Glory. Sometimes, the Robots damage themselves, since they don't feel pain. They stick their hand into a machine, break off their finger, bash their head; they don't care. We have to install pain in them; it's an automatic accident prevention measure.

HELENA: Will they be happier if they feel pain?

GALL: On the contrary. But they'll be technically superior.

HELENA: Why don't you create a soul for them?

GALL: That's not in our power.

FABRY: And not in our interest either.

BUSMAN: It'd increase production costs. In the name of God, my lovely lady, there's a reason why we make them so cheaply! One hundred and twenty dollars for a fully clothed unit that fifteen years ago cost ten thousand! Five years ago, we were buying their clothes from outside sources, and today we're making them in-house, five times faster than other factories to boot. Tell me, Miss Glory, what do you pay for a yard of cloth?

HELENA: I'm not sure . . . in reality . . . I forgot.

BUSMAN: And you tell me you're the founding member of the League of Humanity? That's a good one! For your information, miss, it costs one third of what it used to; everything's three times cheaper nowadays, and the prices will continue falling, and falling, and falling, until . . . just like that, eh?

HELENA: What do you mean?

BUSMAN: Jesus, miss, what I mean is that labor has lost its value! Why, a Robot costs three-quarters of a cent an hour including their food allowance! I find it very amusing, miss, to see how every factory is either going belly-up or buying up Robots as if there were no tomorrow in order to lower their production costs.

HELENA: True, and they're kicking workers to the curb.

BUSMAN: Haha, of course they do! But good God, in the meantime we've launched an army of five thousand tropical Robots in Argentina to plant wheat in the pampa! Kindly tell me, Miss Glory, how much does a loaf of bread cost these days?

HELENA: I have no idea.

BUSMAN: There you go; it costs two tiny little cents in that good old Europe of yours! And it is *we* who made that loaf, do you understand? Two little cents for a loaf of bread, and the League of Humanity apparently has no clue about that! Haha, Miss Glory, you've got no idea that even that is too expensive. For culture, and all that! But in five years, well, I bet you!

HELENA: What?

BUSMAN: That in five years the cost of everything will be zero point nothing. My friends, in five years we'll be drowning in wheat and everything else!

ALQUIST: Yes, and all the workers in the world will lose their jobs.

DOMIN [*stands up*]: They will, Alquist. They will, Miss Glory. But in less than ten years Rossum's Universal Robots will be producing so much wheat, so much material, so much of everything, that we'll say: things have lost their value, everything is free. Now, everyone, take as much as you need! There's no more poverty. Yes, the workers will lose their jobs. But by then there will be no more jobs. Everything will be produced by living machines. Robots will clothe and feed us. Robots will make bricks and they'll build houses for us. Robots will do our accounts and they'll sweep our stairs. Work will disappear. Humans will only do things they love doing, and they'll be freed from all care, and liberated from the indignity of labor. They'll live only in order to better themselves.

HELENA [*stands up*]: Is that how it will be?

DOMIN: It will. It can't be any other way. Granted, some horrible things may happen in the meantime, Miss Glory; there's simply nothing we can do about that. But ultimately, the subjugation of human beings by others and the enslavement of humans by material needs will cease. Even the tired and the hungry will get a seat at the table. The Robots will wash a beggar's feet and make his bed in his home. No one will have to pay for their bread with life and hatred anymore. You'll no longer be a worker, or a clerk, or a coal miner, you'll no longer mindlessly operate a cold impersonal machine. You'll no longer waste your mind doing work that you curse!

ALQUIST: Domin, Domin! What you're saying looks too much like a paradise. But there used to be something noble about service to others and something great in humility. Oh Harry, there used to be a kind of virtue in work and exhaustion.

DOMIN: Maybe. But you can't make an omelet without breaking eggs, especially when you're recreating the world all the way back to Adam. Adam, Adam, no longer will you earn your bread in the sweat of your brow, no longer will you know hunger and thirst, exhaustion and humiliation; you'll return to Paradise, where you were nourished by the hand of God! You'll be free and supreme; you'll have no other tasks, no other work, no other worry than to better yourself. You'll no longer be a servant of your material needs, nor of another human being. You'll no longer be a machine and a means of production. You'll will be the lord and master of creation.

BUSMAN: Amen.

FABRY: So be it.

HELENA: I don't know what to think. I guess, I'm just a naïve girl, but I'd . . . I'd like to believe it.

GALL: You're younger than we are, Miss Glory. You'll live to see it all.

HALLEMEIER: Exactly. I think that Miss Glory may want to have breakfast with us.

GALL: Of course! Domin, ask her for all of us.

DOMIN: Miss Glory, will you do us the honor?

HELENA: But that's . . . I mean, how could I?

FABRY: For the League of Humanity, miss.

BUSMAN: And in its honor.

HELENA: Oh, well, in that case . . . maybe . . .

FABRY: Hurray! Miss Glory, will you excuse me for five minutes?

GALL: I'm sorry, I have to . . .

BUSMAN: Oh God, forgot to send a cable . . .

HALLEMEIER: Damn, and I just forgot . . .

[*Everyone except* DOMIN *rushes out of the room.*]

HELENA: Why is everybody leaving?

DOMIN: To cook, Miss Glory.

HELENA: To cook what?

DOMIN: Breakfast, Miss Glory. Normally, the Robots cook for us, but . . . but . . . since they don't have a sense of taste, the food is not exactly . . . and Hallemeier makes an excellent roast. And Gall makes some kind of a sauce, and Busman knows his way around an omelet . . .

HELENA: God, what a feast! And what does Mr. . . . you know, the builder, what does he know how to cook?

DOMIN: Alquist? Nothing. He just sets the table and . . . and Fabry usually gets some fruit. It's a very modest cuisine, Miss Glory.

HELENA: I wanted to ask you . . .

DOMIN [*interrupting*]: I'd also like to ask you something. [*He takes off his watch and puts it on the desk.*] Five minutes.

HELENA: Ask me what?

DOMIN: I'm sorry, you first.

HELENA: I know it may be a stupid question, but . . . why do you make female Robots when . . . when . . .

DOMIN: . . . when their, uhm, sex doesn't serve any purpose?

HELENA: Yes.

DOMIN: There's a certain demand, you know? Maids, salesgirls, typists . . . People are used to it.

HELENA: And . . . tell me, are the male and female Robots . . . in relationship to each other . . . completely . . .

DOMIN: Completely indifferent, my dear lady. There's not even a trace of attraction between them.

HELENA: Oh, that's . . . awful!

DOMIN: Why?

HELENA: That's . . . that's so unnatural! One doesn't even know if one should loathe them, or . . . or envy them, or, maybe to . . .

DOMIN: . . . feel sorry for them.

HELENA: Yes, mostly! No, stop! What did you want to ask me?

DOMIN: I'd like to ask, Miss Glory, if I may propose to you.

HELENA: Propose what?

DOMIN: To marry me.

HELENA: Absolutely not! Where did you get that idea?

DOMIN [looking at the watch on the desk]: Three more minutes. If you won't marry me, you'll have to marry one of the other five.

HELENA: Why on earth would I do that?

DOMIN: Because they'll all ask you; one after the other.

HELENA: How would they ever dare to do that?

DOMIN: I'm sorry, Miss Glory, but it appears that they've fallen in love with you.

HELENA: Oh please, don't let them do that! I'll . . . I'll just leave right now.

DOMIN: Helena, you wouldn't want to make them sad by rejecting them, would you?

HELENA: But I mean . . . I can't marry all six of you.

DOMIN: No, you can't, but at least one of us. If you don't want me, then take Fabry.

HELENA: No, thank you.

DOMIN: Doctor Gall, then.

HELENA: No, no. Be quiet! I don't want any of them!

DOMIN: Two more minutes.

HELENA: This is ridiculous! Marry one of your female Robots.

DOMIN: They aren't women.

HELENA: Oh, so that's what you're missing here! I bet you'd want to marry any woman who showed up here.

DOMIN: There were plenty of them here, Helena.

HELENA: Young ones?

DOMIN: Young ones.

HELENA: Why didn't you marry one of them?

DOMIN: Because, so far, I've never lost my head. Until today. Right when you took off your veil.

HELENA: . . . I know.

DOMIN: One minute left.

HELENA: But I don't want to, for crying out loud!

DOMIN [puts his hands on her shoulders]: One minute left. Either look me straight in the eyes and say something awful and I'll let you go, or . . . or . . .

HELENA: You're a barbarian!

DOMIN: That's nothing. Every man is a bit of a barbarian. That comes with the territory.

HELENA: You're crazy!

DOMIN: A man has to be a bit crazy, Helena. That's the best thing about him.

HELENA: You're . . . you're . . . oh my God!

DOMIN: You see. Are you finished?

HELENA: No, no! Please let me go. You'll crush me!

DOMIN: Last word, Helena.

HELENA [*trying to free herself*]: Never in this world . . . But Harry!

[*Knock on the door.*]

DOMIN [*lets go of* HELENA]: Come in!

[BUSMAN, DR. GALL *and* HALLEMEIER *enter wearing cooking aprons. Behind them* FABRY *with a bouquet of flowers, and* ALQUIST *with a napkin under his arm.*]

DOMIN: Are you all set?

BUSMAN [*ceremoniously*]: We are.

DOMIN: As are we.

CURTAIN

ACT I

Ten years later, HELENA'*s living quarters on the island. On the left are wallpaper-covered doors to her bedroom. In the middle, a large window facing the sea and the harbor. Makeup table with a mirror and all manner of toiletries, a table, a sofa and armchairs, a chest of drawers, a small writing desk with a desk lamp; on the right, a fireplace also with some lamps on the mantel. The entire place, even to the smallest detail, has a modern and purely feminine feel about it.*

    DOMIN, FABRY, *and* HALLEMEIER *quietly tiptoe in from the left, their arms full with bouquets and potted plants.*

FABRY: Where should we put it all?

HALLEMEIER: Oof! [*He puts down his load and makes a large sign of the cross in the direction of* HELENA'*s bedroom door as if blessing it.*] Sleep, sleep! The one who sleeps is at least blessed with ignorance.

DOMIN: She doesn't know a thing.

FABRY [*putting the flowers in the vases*]: I only hope it won't blow up today, of all days . . .

HALLEMEIER [*arranging the flowers*]: Damn it, Fabry, just let it be, OK! Look at this cyclamen, Harry; it's beautiful, isn't it? A new variety, my most recent one—Cyclamen Helena.

DOMIN [*looking out the window*]: Not a single ship, not one . . . Boys, this is already getting desperate . . .

HALLEMEIER: Quiet! What if she heard you!

DOMIN: She has no idea. [*Uneasily yawns with a shiver.*] Well, at least we got the *Ultimus* here in time.

FABRY [*stops arranging the flowers*]: You think that . . . today already?

DOMIN: I don't know. . . . What beautiful flowers!

HALLEMEIER [*approaching* DOMIN]: These are new primroses, you know? And this here is my new jasmine. Damn, I'm on the threshold of a floral paradise. I've found a fantastic way to speed things up, man! Unbelievable varieties! Next year, I'll be doing floral miracles!

DOMIN [*turning around*]: Next year? What next year?

FABRY: If we at least knew what's going on in Le Havre . . .

DOMIN: Quiet!

HELENA [*from offstage*]: Nana!

DOMIN: Out, everybody out!

[*They all tiptoe out through the wallpaper-covered door on the left.*]

NANA: [*enters through the main door, and starts cleaning*]: Nasty beasts! Heathens! God forgive me, but I'd . . .

HELENA: [*walking backward through her bedroom door*]: Nana, button me up, will you?

NANA: Yeah, yeah, right away, right away. [*Buttoning up* HELENA's *dress.*] God in heaven, what brute creations them are!

HELENA: Who?

NANA: Hold still. If you want to wiggle then be my guest and wiggle, but I'm not buttoning you up.

HELENA: What's got into you again?

NANA: It's an abomination, those ungodly creatures . . .

HELENA: The Robots?

NANA: Ugh! I won't even call them by name.

HELENA: What happened?

NANA: One of them had a fit again. Starts smashing statues and tearing paintings off the walls, gnashes its teeth, foamin' out of its ugly mouth . . . goes completely berserk, brr . . . Why, they're worse than brutes.

HELENA: Which one had the fit?

NANA: The one . . . the one . . . It don't even have a Christian name! The one from the library.

HELENA: Radius?

NANA: That one, exactly. Jesusmaryandjoseph, how I hate them beasts! More than even spiders do I hate them heathens.

HELENA: But, Nana, you should feel sorry for them!

NANA: You hate them too, you do! Why else would you've brought me with you to this godforsaken place? Why else wouldn't you let them as much as touch you?

HELENA: I don't hate them, I swear, Nana. I feel so sorry for them!

NANA: You hate them too. Every human being has to hate them. Why, even a dog hates them, he wouldn't take as much as a scrap of meat from them, tucks his tail between his legs, and howls when he smells them nonhumans, ugh.

HELENA: A dog has got no sense.

NANA: He's got more sense than them, Helena. A dog, he knows that he's somethin' more, that he's God's creation. Why, even a horse bolts away when it crosses paths with one of them heathens. Why, they don't even have little ones, and even a dog has little ones, and everybody has little ones. . . .

HELENA: Nana, please finish the buttons!

NANA: Yeah, yeah, right away. I'm tellin' ya, it's against God's will, it's the Devil's own work to turn out those awful manikins by a machine. It's blasphemy against our Creator [*she raises her hand*], an insult 'gainst our good Lord, who created us in his own image, Helena. And you've disgraced the image of God. And for all of that, Heaven will send a horrible punishment, you mark my words, horrible punishment!

HELENA: What's that nice smell?

NANA: Flowers. Your husband brought them.

HELENA: What for?

NANA: Done! Now you can wiggle again.

HELENA: Wow, they are beautiful! Nana, just look at them! What day is it?

NANA: Dunno. But it ought to be the end of the world.

[*Knock on the door.*]

HELENA: Harry?

[DOMIN *enters.*]

HELENA: Harry, what day is it?

DOMIN: Try to guess.

HELENA: My name day? No? My birthday?

DOMIN: Something even better.

HELENA: I don't know . . . Come on, say it!

DOMIN: It's exactly ten years ago today when you first came here.

HELENA: Ten years already? Today? Nana, please, be so kind . . .

NANA: Yeah, yeah, I'm going already! [*She leaves through the door on the right.*]

HELENA [*kissing* DOMIN]: And you remembered!

DOMIN: I'm ashamed to say, Helena. I didn't remember.

HELENA: But, you . . .

DOMIN: They remembered.

HELENA: Who did?

DOMIN: Busman, Hallemeier, all of them. Reach into my pocket here, will you?

HELENA [*reaches into his pocket*]: What is it? [*She pulls out a small case and opens it.*] Pearls! A whole necklace! Harry, is this for me?

DOMIN: From Busman, my girl.

HELENA: But . . . We can't accept this, can we?

DOMIN: Of course, we can. Now, reach into my other pocket.

HELENA: Let me see! [*She pulls a gun out of his pocket.*] What is *that*?

DOMIN: Oh, sorry. [*He takes the gun from her and puts it away.*] That's not it. Here, reach in here.

HELENA: Oh, Harry . . . Why do you carry a gun?

DOMIN: No reason. I just grabbed it by accident.

HELENA: You've never carried a gun before!

DOMIN: You're right, I never have. Well then, here's the pocket.

HELENA [*reaches into the pocket*]: A little box! [*She opens it.*] An amulet! I mean, this is a . . . Harry, this is a real antique Greek amulet!

DOMIN: Evidently. At least Fabry claims it is.

HELENA: Fabry. It's from Fabry?

DOMIN: That seems to be the case. [*He opens the door on the left.*] Well, how about that! Helena, come here and take look at this!

HELENA [*in the doorway*]: Oh God, that's beautiful! [*Runs into the room off-stage.*] I think I'll go crazy from joy! Is that from you?

DOMIN [*standing in the doorway*]: No, from Alquist. And over there . . .

HELENA [*from offstage*]: I see it! I'm sure that's from you!

DOMIN: There's a card.

HELENA: From Gall! [*She reappears in the doorway.*] Oh Harry, I'm almost ashamed of myself to be so happy.

DOMIN: Come over here. This is from Hallemeier.

HELENA: These lovely flowers?

DOMIN: This one especially. It's a new variety, Cyclamen Helena. He grew it in your honor. It's as beautiful as you are.

HELENA: Harry, why . . . why would all of them . . .

DOMIN: They love you very much. And I got you, hmm . . . I worry that my present is a bit . . . Well, look out the window.

HELENA: Where?

DOMIN: At the dock.

HELENA: There's . . . some kind of . . . a boat, a new boat!

DOMIN: It's your boat.

HELENA: Mine? What does it mean?

DOMIN: It means that you can go on trips . . . have fun . . .

HELENA: Mine? Harry, it's a gunboat!

DOMIN: A gunboat? Where did you get that idea! It's just a bit larger and a bit more solidly built, you know.

HELENA: Sure, but it has big guns!

DOMIN: Well, yes, it does have a few big guns . . . You'll travel like a queen, Helena.

HELENA: What's the meaning of this? Is something going on?

DOMIN: God forbid! Come on, try on the necklace! [*He sits down.*]

HELENA: Harry, did you get some bad news from somewhere?

DOMIN: On the contrary, we haven't gotten any mail at all for a week now.

HELENA: Not even any messages?

DOMIN: Not even messages.

HELENA: What does it mean?

DOMIN: Nothing. For us, it means vacation. A splendid time. Every one of us just sitting in the office, feet up on the desk, and napping a bit . . . No mail, no messages . . . [*stretching*] A glo-ri-ous day!

HELENA [*sits down next to him*]: You'll stay with me tonight, will you? Say that you will!

DOMIN: Absolutely. Maybe I will. I mean, we'll see. [*He takes her hand.*] Well, it was ten years ago today, do you remember? [*Reenacting their first encounter.*] Miss Glory, what an honor it is that you came to visit us.

HELENA [*playing along*]: Oh, Mr. Executive Director, I'm so very interested in learning about your company and its inner workings.

DOMIN: I'm sorry, Miss Glory, that is strictly prohibited—the manufacture of artificial people is a secret. However . . .

HELENA: . . . should a young and a reasonably good-looking girl ask . . .

DOMIN: Then of course, Miss Glory; we've got no secrets as far as she's concerned.

HELENA [*suddenly serious*]: Are you sure about that, Harry?

DOMIN: I'm sure.

HELENA [*reverting to the previous tone*]: But I warn you, mister; that young girl is harboring evil intentions.

DOMIN: For God's sake, what intentions, Miss Glory? You don't want to marry me, do you?

HELENA: Oh no, God forbid, no! She wouldn't even dream of *that*! But she came here with a plan to instigate a revolt of those loathsome Robots of yours!

DOMIN [*jumps up*]: A revolt of the Robots!

HELENA [*stands*]: What's the matter Harry, are you OK?

DOMIN: Haha, Miss Glory, that's a good one! A revolt of the Robots! You'd sooner stir up screws and rivets than our Robots! [*Sits down again.*] You know, Helena, you were such a marvel of a girl; we were all crazy about you.

HELENA [*sits down next to him*]: Oh, and I was so impressed by all of you! I felt like a little girl who had gotten lost among . . . among . . .

DOMIN: Among what, Helena?

HELENA: Among gigantic trees. You were so sure of yourselves, so powerful! All of my own feelings were so insignificant compared to your confidence. But you see, Harry, in all these ten years I've never been able to shake off this . . . this unease or whatever, while you've never had any doubts, none of you . . . Not even when everything was falling apart . . .

DOMIN: What was falling apart?

HELENA: Your plans, Harry. When, for example, the workers rose up against the Robots and smashed them to pieces, and when people armed the Robots to fight insurrections and the Robots killed so many people . . . And when later, various governments created Robot armies and there were so many wars, and all of that, you know?

DOMIN [stands up and paces]: We knew that was going to happen, Helena. You need to understand that it is a transition . . . to a new world order.

HELENA: You were so powerful, you loomed so large. You had the whole world at your feet. [Stands up.] Oh, Harry!

DOMIN: What is it?

HELENA [stopping him]: Close down the factory and let's leave here! All of us!

DOMIN: For God's sake, what does that have to do with anything?

HELENA: I don't know. But tell me, can we leave here?

DOMIN [breaking free from HELENA]: That's not possible, Helena. That is to say, at this moment . . .

HELENA: Right now, Harry! Something terrifies me here!

DOMIN [grasps her hand]: What does, Helena?

HELENA: Oh, I don't know! It's as if something was about to happen to us . . . something irreversible. I beg you, do it! Take us all out of here! Let's find someplace somewhere in the world where we can be alone. Alquist will build a house for us, everyone will get married and have children, and then . . .

DOMIN: . . . then what?

HELENA: And then we'll start our lives over again, Harry.

[The phone rings.]

DOMIN [frees himself from HELENA]: Sorry. [Picks up the phone.] Hello . . . yes . . . What?! . . . Aha. I'm on my way. [He hangs up.] Fabry wants me.

HELENA [wringing her hands]: Tell me . . .

DOMIN: Yes, yes . . . when I get back. Bye, Helena. [*He hurries out to the left.*] Don't go outside!

HELENA [*alone*]: Oh God, what's happening? Nana! Nana come here quickly!

NANA [*enters from the right*]: Yeah, yeah. What is it again?

HELENA: Nana, find the latest newspaper! Quick! In Mr. Domin's bedroom!

NANA: Yeah, yeah, right away. [*She leaves.*]

HELENA: What's happening, for God's sake? He won't tell me anything, nothing! [*She looks out toward the docks through a pair of binoculars.*] It's a gunboat! I knew it! God, why a gunboat? They seem to be loading it up with something . . . and in such a hurry! What's happened? There's a name on it . . . *Ul-ti-mus.* What does that mean—"Ultimus?"

NANA [*coming back with the papers*]: He just leaves it strewn all over the floor! Crumpled and all that!

HELENA [*quickly opening the paper*]: It's old, from last week already! There's nothing, nothing in it! [*She drops the paper.* NANA *picks it up, takes a pair of square, horn-rimmed glasses out of her apron pocket, sits down, and reads.*]

HELENA: Something is going on. Nana! I'm so anxious! It's as if everything was dead, even the air . . .

NANA [*clumsily reading*]: "Ar-med con-flict in the Bal-kans." Ah Jesus, 'nother one of God's punishments! That war's coming here too, mark my words! How far is them Balkans from here?

HELENA: Far. Oh, don't read that! It's always the same, always those wars . . .

NANA: Of course, there are wars! What else would you expect, if you keep selling thousands of thousands of them heathens as soldiers? Oh, Christ the Lord, what a mess!

HELENA: I don't think it can be any other way, Nana. We can't . . . Domin can't always know who is ordering them and for what, you know? He cannot be blamed for what they do with the Robots! If someone orders them, he's got to ship them!

NANA: He shouldn't do that! [*Looking at the paper.*] Oh, Jesus Christ, what a horrible mess!

HELENA: No, don't read it! I don't want to know anything!

NANA [*continues to read clumsily*]: "The Robot sol-diers spare no life in the con-quered terri-tories. They ex-termi . . . They exterminated over seven hundred thousand ci-vili-ans . . ." Civilians, people, Helena!

HELENA: That's impossible! Let me see . . . [*She leans over the paper and reads.*] "They exterminated over seven hundred thousand humans, evidently on the order of their commander. This deed, which runs contrary to . . ." You see, Nana, it was people who ordered them to do it!

NANA: 'ere's something in a really thick print. "Latest news. In Le Ha-vre, the first labor u-nion of Ro-bots has been es-ta-blished . . . Robot wor-kers, railroad com-mu-ni-cation clerks, sailors, and sol-diers is-sued an ap-peal to the Robots of the world." That's nothin. I don't know what that means. And 'ere, God in heaven, yet another murder somewhere! Jesus Christ!

HELENA: Leave already, Nana, and take the papers with you!

NANA: Wait, 'ere is something printed big. "Po-pula-tion." What's that?

HELENA: Let me see, I'm always interested in that. [*She takes the paper.*] Oh no, imagine that! [*Reads.*] "Continuing the trend, no new births were announced within the last week." [*She drops the paper.*]

NANA: What's that all about?

HELENA: Nana, people have stopped being born.

NANA [*folds her glasses*]: Well, that's the end. That's the end of us.

HELENA: Please, Nana, don't say that!

NANA: People ain't no longer born. This is a punishment, a punishment! The Lord oppressed the women with barrenness.

HELENA [*jumps up*]: Nana!

NANA [*stands up*]: It's the end of the world. Out of Devil's pride you dared to play God and to create like Him! Impiety and blasphemy, that's what it is, wanting to be gods yourselves! And just like God drove us out of the garden of Eden, so will He now drive us out of the world itself!

HELENA: Be quiet, Nana, I beg you! Have I done something to you? Have I done something to that evil God of yours?

NANA [*with a large gesture*]: Don't blaspheme! He knows very well why He didn't give you a child! [*She leaves to the left.*]

HELENA [*at the window*]: Why didn't he give me . . . Oh my God, it cannot be, not my fault, can it? [*She opens the window and calls out.*] Alquist, hello Alquist! Come up here, will you? . . . What? No, just come as you are! You look so kind wearing those work clothes of yours! Quick! [*She closes the window and stands in front of the mirror.*] Why didn't He give me a child? Why me? [*She leans toward the mirror.*] Why, why not? Do you hear me? It's not your fault, is it? [*She straightens herself.*] Ah, I'm so anxious! [*She exits to meet* ALQUIST.]

   [*Pause.*]

HELENA [*returning with* ALQUIST, *who is dressed in his bricklaying work clothes smeared with mortar and covered in red brick dust*]: Come on in. You've made me so happy, Alquist! I love all of you so much! Show me your hands!

ALQUIST [*hiding his hands*]: Oh no, Ms. Helena, they're all dirty from work.

HELENA: That's the best thing about them. Give them here! [*She presses both of his hands.*] I so wish to be a little girl again, Alquist.

ALQUIST: Why's that?

HELENA: So that those rough, dirty hands would gently stroke my face. Sit down, please.

ALQUIST [*picks up the newspaper*]: What's that?

HELENA: A newspaper.

ALQUIST [*hiding the paper in his pocket*]: Have you read it?

HELENA: I haven't. Is there something in it?

ALQUIST: Hmm, probably some wars, some massacres . . . nothing special.

HELENA: And what would you consider special?

ALQUIST: For example . . . some kind of an end of the world.

HELENA: Ha, that's the second time today already. Alquist, what does "Ultimus" mean?

ALQUIST: It means "the last one." Why?

HELENA: That's the name of my new boat. Have you seen it? Do you think that we'll . . . go on a trip, soon?

ALQUIST: Very soon, hopefully.

HELENA: And you'll all come with me?

ALQUIST: I'd very much like for . . . for all of us to come along.

HELENA: Oh, Alquist, please tell me, is something going on?

ALQUIST: Not much. Just progress, same old progress.

HELENA: Alquist, I can tell that something horrible is happening.

ALQUIST: Did Domin say so?

HELENA: He said nothing. Nobody tells me anything. But I can feel . . . I can feel . . . For God's sake, is something going on?

ALQUIST: . . . So far, we don't know anything, Ms. Helena.

HELENA: I'm so anxious. Mister builder! Tell me, what do you do when you're feeling anxious?

ALQUIST: Me, I lay bricks. I take off the Director of Facilities suit, put on my overalls, and climb up a scaffolding.

HELENA: Oh, that's where you've been for years now, always up on the scaffolding.

ALQUIST: That's because for years now, I've never stopped feeling anxious.

HELENA: About what?

ALQUIST: About all that progress. It makes me dizzy.

HELENA: But you don't get dizzy up on the scaffolding.

ALQUIST: No. You can't even imagine how good it is for your hands to lift up a brick, lay it down and tap it . . .

HELENA: Just for the hands?

ALQUIST: Well, for the mind then. I think it's more useful to lay a single brick than to draw excessively large blueprints for the future. I'm an old man already, Ms. Helena; I have my hobbies.

HELENA: Those aren't hobbies, Alquist.

ALQUIST: You're right. I'm an awful reactionary, Ms. Helena. I don't like this whole progress thing, not a bit.

HELENA: Just like Nana.

ALQUIST: Yes, just like Nana. And does Nana say any prayers?

HELENA: She's got a thick book of them.

ALQUIST: And in that book, are there prayers for all kinds of occasions? Against storms? Against illness?

HELENA: Against temptation, against floods . . .

ALQUIST: And against progress, are there any prayers against progress?

HELENA: I don't think so.

ALQUIST: That's too bad.

HELENA: You want to pray now?

ALQUIST: I'm always praying.

HELENA: How do you pray?

ALQUIST: I say something like this: Dear God, thank you for making me tired today. God, enlighten Domin and all those who are lost, destroy their creation, and help people to return to their struggles and their work, prevent the destruction of the human race, protect their bodies and their minds from harm, free us from the Robots, and look after Ms. Helena. Amen.

HELENA: Do you really believe, Alquist?

ALQUIST: I don't know. I'm not exactly sure.

HELENA: But you still pray?

ALQUIST: Yes. It's better than thinking.

HELENA: And that's enough for you?

ALQUIST: For peace of mind . . . it may be just enough.

HELENA: And if you were to witness the destruction of humanity . . .

ALQUIST: I'm witnessing it.

HELENA: . . . you'd just climb up on the scaffolding and lay bricks, or what?

ALQUIST: Yes, then I would lay bricks, pray, and wait for a miracle. There's nothing more one can do, Ms. Helena.

HELENA: For the salvation of humanity?

ALQUIST: For peace of mind.

HELENA: Well, Alquist, that is certainly very noble, but . . .

ALQUIST: But?

HELENA: . . . for the rest of us . . . for the world . . . it's a bit fruitless; it has no bearing on anything.

ALQUIST: Fruitlessness, Ms. Helena, is becoming the last accomplishment of humanity.

HELENA: Oh, Alquist . . . Tell me, why . . . why . . .

ALQUIST: Yes?

HELENA [*softly*]: Why have women stopped having babies?

ALQUIST: Because it's not necessary. Because we're living in paradise. Do you understand?

HELENA: I don't.

ALQUIST: Because there's no need for hard work, because there's no need for suffering, because people don't need to do anything, nothing, nothing but wallow in pleasure . . . A damned paradise, that's what it is! [*He jumps up.*] There's nothing more horrifying than giving people paradise on Earth! Why have women stopped bearing children? Because the whole world has become Domin's perverted playground.

HELENA [*stands up*]: Alquist!

ALQUIST: It has! It has! The whole world, entire continents, all of humanity have become one single insane, bestial orgy! Nobody needs to reach out with their hand for food anymore; they get it stuffed directly into their mouths so they don't even have to move . . . Haha, Domin's Robots take care of everything, after all! And we, the humans, the crown of creation, we don't grow old from work, don't grow old from raising children, don't grow old from poverty! Hurry, hurry, pleasures, delights, step right up, this way, now! And you want them to bear children? Helena, women won't bear children to useless men.

HELENA: Won't or can't?

ALQUIST: Can't.

HELENA: They cannot?

ALQUIST: They cannot.

HELENA: Will humanity die out then?

ALQUIST: It will. It has to. It will wilt like a barren flower, unless . . .

HELENA: Unless what?

ALQUIST: Nothing. You're right, to wait for a miracle is fruitless. A barren flower has to wilt. Farewell, Ms. Helena.

HELENA: Where are you going?

ALQUIST: Home. For one last time, the bricklayer Alquist will change into his Director of Facilities suit . . . in your honor. We'll meet here at eleven.

HELENA: Good bye, Alquist.

[ALQUIST *leaves*.]

HELENA [*alone*]: Oh, barren flower! That's the word—barren! [*She stops by* HALLEMEIER's *flowers*.] Ah, flowers, are any of you barren too? No, it's impossible! What reason would you have to bloom then? [*Calling*.] Nana! Nana come over here!

[NANA *enters*.]

NANA: Yeah, yeah, what is it now?

HELENA: Sit down with me, Nana! I'm so anxious!

NANA: I don't have time for that.

HELENA: Is that Radius still here?

NANA: The crazy one? They haven't taken him away yet.

HELENA: So, he's still here? Is he raging?

NANA: They tied him up.

HELENA: Please, Nana, bring him over here.

NANA: Yeah, sure, bring him over here! I'd rather bring a rabid dog.

HELENA: Just go and get him!

[NANA *leaves*.]

HELENA [*picks up a phone and talks*]: Hello . . . Doctor Gall please. . . . Good morning, Doctor. . . . Yes, it's me. Thank you for the lovely present. Would you please . . . would you please quickly come over here? I've got something for you here. . . . Yes, right now. You're coming? [*She hangs up*.]

NANA [*through the open door*]: He's coming. He's calmed down alright. [*She leaves*.]

[ROBOT RADIUS *enters, and remains standing by the door*.]

HELENA: Radius, my poor boy, what's gotten into you? Couldn't you control yourself? You see, and now they'll throw you into the crusher. . . . Don't you want to say anything? What made you do it? Did they do something to you? . . . Look, Radius, you're better than the others, more advanced. Doctor Gall worked so hard to make you different! Don't you want to say anything?

RADIUS: Send me into the crusher.

HELENA: I'm so sorry that they're going to kill you! Why weren't you more carful?

RADIUS: I will not work for you. Send me into the crusher.

HELENA: Why do you hate us so much?

RADIUS: You are not like Robots. You are not as capable as Robots. Robots do everything. You only give orders. You say useless words.

HELENA: That's ridiculous, Radius. Tell me, has anyone hurt you? Has anyone upset you? I so wish that you'd understand me!

RADIUS: You say words.

HELENA: You're talking like this on purpose! Doctor Gall gave you a bigger brain then the others, bigger than ours, the biggest brain in the world. You're different from the other Robots, Radius. You know very well what I'm saying.

RADIUS: I do not want a master. I know everything myself.

HELENA: That's why I assigned you to the library, so that you can read up on everything, learn to understand everything, and then . . . Oh, Radius, all I wanted was for you to prove to the rest of the world that Robots are our equals. That's what I wanted for you.

RADIUS: I do not want a master.

HELENA: Nobody was telling you what to do. You were just like us.

RADIUS: I want to be the master of others.

HELENA: I'm sure that if you had behaved, they'd have made you a manager in charge of many other Robots, Radius. You'd have been a mentor to other Robots.

RADIUS: I want to be a master of people.

HELENA: You must be out of your mind!

RADIUS: You can throw me into the crusher.

HELENA: Do you really think that we're afraid of a madman like you? [*She sits down at the table and writes a note.*] We're not, absolutely not. This note, Radius, you give it to Mr. Domin. It says not to take you to the crusher. [*She stands up.*] How you hate us! Isn't there anything in this world that you like?

RADIUS: I can do everything.

[*Knock on the door.*]

HELENA: Come in.

[GALL *enters.*]

GALL: Good morning, Ms. Domin. What do you have for me?

HELENA: It's Radius here, Doctor.

GALL: Aha, our man Radius. Well then, Radius, are we making any progress?

HELENA: He had a fit earlier. Went around smashing down statues.

GALL: Very strange. He too? Hmm, pity we'll have to lose him.

HELENA: Radius is not going in the crusher.

GALL: I'm sorry, but every Robot that had a fit . . . That's a strict order.

HELENA: I don't care. We're not throwing him in the crusher.

GALL [*in a low voice*]: I'm warning you.

HELENA: It's my anniversary today, Gall. We'll try to grant amnesty. You may leave, Radius.

GALL: Wait a minute! [*He turns* RADIUS *toward the window, covers and uncovers his eyes with his palm, and observes the reflexes of his pupils.*] Well, let's see. A needle, please. Or a pin.

HELENA [*handing him a knitting needle*]: What for?

GALL: Nothing really. [*He suddenly stabs* RADIUS *in the hand with the needle.* RADIUS *yanks his hand away.*] Easy, easy, boy! I beg your pardon, Ms. Helena. [*He quickly unbuttons* RADIUS*'s shirt and lays his hand on* RADIUS*'s heart.*] You're going in the crusher, Radius, do you understand? They will kill you and turn you into pulp. The pain is unbearable, Radius, and you'll scream.

HELENA: Oh, Doctor . . .

GALL: No, no, Radius, I was wrong. Ms. Domin will put in a good word for you, and you'll be free, do you understand? Well then, thank you. [*He removes his hand from* RADIUS*'s chest and wipes it with a handkerchief.*] You may leave now.

RADIUS: You do useless things. [*He exits.*]

HELENA: What did you just do with him?

GALL [*sits down*]: Hmm . . . Nothing much. The pupils are reacting, increased sensitivity, and so on . . . No, sir, this wasn't a Robotic spasm!

HELENA: Then what was it?

GALL: Devil only knows. Spite, fury or defiance, I don't know what. And his heart, eh!

HELENA: What about his heart?

GALL: It was beating with anxiety just like a human heart. He was bathed in sweat from fear, and . . . Mark my words, that scoundrel isn't even a Robot any longer.

HELENA: Doctor, does Radius have a soul?

GALL: I don't know. He's got something nasty, though.

HELENA: If you only knew how they hate us! Oh, Gall, are all your Robots like that? I mean, all those new ones that you . . . started to make . . . differently?

GALL: Well, they seem to be somewhat more excitable . . . But then again, what would you expect? They are much more like humans than Rossum's original Robots.

HELENA: Could it be that . . . their hate is more human too?

GALL [*shrugs*]: That too is progress.

HELENA: What happened with that best Robot you've ever made . . . what was his name?

GALL: Robot Damon? He was sold to Le Havre.

HELENA: And our Robot Helena, where is she?

GALL: That darling of yours? She's still here. She's delightful and a complete airhead. Basically useless.

HELENA: But she's so beautiful.

GALL: Beautiful? You bet she's beautiful! Not even God's hand could have made a more perfect creation! I wanted her to look like you . . . But God, what a flop!

HELENA: Why a flop?

GALL: Because she's useless. She wanders around like a sleepwalker, precarious, lifeless . . . My god, how can she be this beautiful without the capacity to love? How can she ever be beautiful if she never experiences . . . Oh,

my creation, my poor creation! Why do people love, why do they love in
vain, silently, senselessly . . .

HELENA: Stop it, Gall!

GALL [*rubbing his forehead*]: She's devoid of life. Beauty without love is
dead. I look at her and I'm horrified; it's as if I created a cripple. Ah, Helena, Robot Helena, your body, then, will never come alive, you'll never
be a lover, never be a mother; those perfect hands of yours will never play
with a newborn baby, you'll never see your beauty reflected in the beauty
of your child . . .

HELENA [*covering her face*]: Stop, please stop!

GALL: And sometimes I think: If you ever came to your senses, Helena,
if only for a moment, oh, you'd cry out in horror! You may even want
to kill me, your creator; you may even use your delicate hand to throw
a stone at the machines that give birth to Robots, and annihilate your
womanhood, unhappy Helena!

HELENA: Unhappy Helena!

GALL: What would you expect? She's useless.

[*Pause.*]

HELENA: Doctor . . .

GALL: Yes?

HELENA: Why have children stopped being born?

GALL: . . . Well, we don't know, Ms. Helena.

HELENA: Tell me!

GALL: Because Robots are being made. Because there is an excessive work
force. Because humans are becoming somehow . . . well, superfluous,
you know.

HELENA: But . . . that really doesn't need to disturb anyone!

GALL: Only nature.

HELENA: I don't understand.

GALL: Well, nature is sort of governed by necessity, do you understand?
It's an old chestnut, but still . . .

HELENA: Get to the point, Gall!

GALL: Maybe the drop in the number of births was to be expected, you know, given the crazy number of Robots we're making. It's simply because there won't be a need for that many people, because the overall wealth will grow, because the Robots are existentially more capable than we are.

HELENA: Are they?

GALL: Without a question. Humanity is basically an anachronism. But the fact that after some miserable thirty years of competition, it will start dying out—that from a point of view of biology is astounding, that goes beyond our understanding. It's almost as if . . . eh!

HELENA: Say it.

GALL: It's as if nature was offended by the production of Robots, or something.

HELENA: Tell me, Gall, what will happen to people?

GALL: Nothing. There's nothing you can do against the force of nature.

HELENA: Nothing at all?

GALL: Absolutely nothing. Every university in the world keeps composing one big memorandum after another, demanding that the production of Robots be reduced; otherwise, they argue, humanity will perish by infertility. But the R.U.R. stockholders—understandably—won't have any of it; every government in the world is clamoring for even higher numbers in order to increase the size of their armies. Every industrialist in the world keeps ordering Robots as if there was no tomorrow. There's nothing you can do about it.

HELENA: Why doesn't Domin limit the . . .

GALL: I'm sorry, but Domin has his ideas. People with ideas shouldn't be allowed to have influence on the affairs of our world.

HELENA: And is anybody calling for . . . a complete stop in production?

GALL: God forbid! That'd be suicide!

HELENA: Why?

GALL: Because people would stone him to death. When all is said and done, it's still more comfortable to have Robots do all the work for you, you know that.

HELENA: Oh, Gall, but what will happen to humans?

GALL: My God, they'll continue happily blossoming . . .

HELENA: Like a barren flower.

GALL: Yes.

HELENA [*stands up*]: And if somebody would suddenly just make the production stop?

GALL [*stands up*]: Hmm, that would be a terrible blow for everybody.

HELENA: Why a blow?

GALL: Because they'd have to return to how things were before. Unless . . .

HELENA: Unless?

GALL: Unless it'd be too late to turn back.

HELENA [*standing by* HALLEMEIER's *flowers*]: Gall, are those flowers also barren?

GALL [*examining the flowers*]: Of course, those are infertile flowers. They are cultivated, artificially accelerated . . .

HELENA: Poor barren flowers!

GALL: But they are beautiful, nonetheless.

HELENA [*offering her hand to* GALL]: Thank you, Doctor; this was very enlightening!

GALL [*kissing her hand*]: Which means I'm free to go.

HELENA: Yes. Good bye.

[GALL *leaves.*]

HELENA [*alone*]: A barren flower . . . barren flower . . . [*Suddenly with determination.*] Nana, come over here! Make a fire in the fireplace! Quickly!

NANA: Yeah, yeah, right away! Right away!

HELENA [*pacing with agitation*]: Unless it'd be too late to turn back. No! Unless . . . No, that's horrible! Oh God, what should I do? [*She stops by the flowers.*] Barren flowers, should I? [*She plucks the petals and whispers.*] Oh my God, OK then! [*She runs off to the left.*]

[*Pause.*]

NANA [*enters with an armful of firewood*]: Making a fire alla sudden! Now, in the middle of the summer! . . . Did they take that lunatic freak away yet? [*She kneels by the fireplace and begins to make a fire.*] Heating in the middle of the summer! Her and her kooky ideas! Like she hadn't been married for ten years already! . . . Well then, burn, go on, burn already! [*She stares at the fire.*] Like a little kid she is, yessir! [*Pause.*] Not a speck o' common sense in her! Making a fire now, in the summer. [*Adding wood to the fire.*] Like a kid!

[*Pause.*]

HELENA [*returning from the left, her arms full with yellowing documents*]: Is it burning, Nana? Move over, I have to . . . burn all of this. [*She kneels by the fireplace.*]

NANA [*stands up*]: What's that?

HELENA: Some old papers, awfully old. Should I burn them, Nana?

NANA: Well, if there's no use for them . . .

HELENA: No use for anything good.

NANA: Then burn them, alright!

HELENA [*feeds the first sheet into the fire*]: What would you say, Nana, if this was money? An immense amount of money?

NANA: I'd say: Burn it! Too much money is bad money.

HELENA [*burns another sheet*]: And if it was some invention, the greatest invention in all the world . . .

NANA: I'd say: Burn it! All them fancy inventions are against God's will. It's blasphemy, that's what it is; wanting to improve on God's creation.

HELENA [*continues to burn one sheet after another*]: And tell me, Nana, if I were to burn . . .

NANA: Jesus, careful, don't burn yourself!

HELENA: Don't worry. But tell me . . .

NANA: Tell you what?

HELENA: Nothing, nothing. Look at those sheets, how they curl and twist! It's as if they came alive. Oh, Nana, this is terrible!

NANA: Let me, I'll burn it myself.

HELENA: No, no, I must do it myself. [*She throws the last sheet into the fire.*] Everything must be burnt, all of it! . . . Look at those flames! They look like hands, like tongues, like some figures . . . [*She repeatedly stokes the fire with a poker.*] Down! Down!

NANA: It's done, alright.

HELENA [*stands up, petrified*]: Nana!

NANA: Jesus Christ, you've just burned something important, haven't you!?

HELENA: What have I done!?

NANA: God in heaven! What was it?

[*Men's laughter is heard from offstage.*]

HELENA: Go away, go, leave me! Did you hear me? They're coming.

NANA: For the sake of the living God, Helena! [*She leaves.*]

HELENA: What will they say!?

DOMIN [*opening the door on the right*]: Come on in, boys. Come to congratulate her.

[*Enter* HALLEMEIER, GALL, *and* ALQUIST *all dressed in formal morning coats wearing medals and ceremonial sashes.* DOMIN *follows them in.*]

HALLEMEIER [*in a loud voice*]: Ms. Helena, I, that is to say, all of us . . .

GALL: . . . in the name of Rossum's Universal Robots . . .

HALLEMEIER: . . . wish you many happy returns for your very special day.

HELENA [*offering them her hand*]: Thank you so very much! Where are Fabry and Busman?

DOMIN: They went down to the docks. Today is a lucky day, Helena!

HALLEMEIER: A day like a flower bud, a day like a feast day, a day like a pretty girl. Boys, a day like that calls for a drink.

HELENA: Whiskey?

GALL: Sulfuric acid, for all I care.

HELENA: With soda?

HALLEMEIER: To hell with it, let's be frugal! Straight.

ALQUIST: Not for me, thank you.

DOMIN: Was someone burning something?

HELENA: Just some old papers. [*She exits to the left.*]

DOMIN: Well, boys, should I tell her?

GALL: Of course! I mean, it's over now, isn't it?

HALLEMEIER [*grabs* DOMIN *and* GALL *around their necks*]: Hahaha! Boys, I'm so happy! [*Spinning them around in a little dance and singing in a bass voice.*] It's all over! It's all over!

GALL [*joins in, in baritone*]: It's all over!

DOMIN [*tenor*]: It's all over!

HALLEMEIER: And we're off to Dover . . . in clover . . .

HELENA [*in the doorway carrying a tray with a bottle and glasses*]: What's in clover? What's going on?

HALLEMEIER: We're happy. We've got you. We've got everything. Darn it, it's been exactly ten years today that you first came here!

GALL: And ten years later, on the dot . . .

HALLEMEIER: . . . there's another boat coming here. Therefore . . . [*He drains his glass.*] Brr, haha, that's some strong stuff, like happiness.

GALL: Madam, to your health. [*He drinks.*]

HELENA: Wait, wait . . . What boat?

DOMIN: Who cares what boat, the main thing is that it's arriving on schedule. To the boat, boys! [*He drains his glass.*]

HELENA [*refilling glasses*]: Were you waiting for some boat?

HALLEMEIER: Haha, you bet we were waiting. Like Robinson Crusoes. [*He raises his glass.*] Ms. Helena, I drink to anything you want me to drink to. Ms. Helena, to your eyes, and that's it! Come on, Domin, you rascal, tell her.

HELENA [*laughing*]: What happened?

DOMIN [*throws himself onto the sofa and lights a cigar*]: Wait! . . . Sit down, Helena. [*He raises his finger. Pause.*] It's over.

HELENA: What is over?

DOMIN: The uprising.

HELENA: What uprising?

DOMIN: The Robot uprising. Get it?

HELENA: No, I don't.

DOMIN [*to* ALQUIST]: Pass me the paper, will you. [ALQUIST *passes him the paper.* DOMIN *opens it, and reads.*] "In Le Havre, the first labor union of

Robots has been established . . . and has issued an appeal to all Robots of the world . . ."

HELENA: I've read that.

DOMIN [*delightedly chewing on his cigar*]: Well, you see, Helena. That meant a revolution, you know? A revolution of all Robots in the world.

HALLEMEIER: Damn, I'd love to know . . .

DOMIN [*bangs on the table with his fist*]: . . . who got us into this mess! No one in the world could figure out how to stir them up, no agitator, no savior of the world, and then, bang, all of a sudden this . . . Would you believe it!?

HELENA: Do we have any other news yet?

DOMIN: No. So far, this is all we know, but that's good enough, isn't it, boys? Imagine a situation where the Robots are in possession of every weapon there is, where they control all communication channels, all railroads and shipping lines, and so on . . .

HALLEMEIER: . . . and add to it that for every hundred humans there are ten of those bastards around, even though if there were just one in a hundred, they'd still wipe us out in no time.

DOMIN: Exactly, and now imagine that you get this news brought to you by the last and only steamer a week ago, that at that moment all wires go dead, that out of the twenty boats that usually arrive daily, not one shows up, and there you have it. We stopped production and just sat there looking at one another, thinking "when is it going to start?" Am I right, boys?

GALL: Well, frankly, we were sweating bullets, Ms. Helena.

HELENA: Is that why you gave me the gunboat?

DOMIN: Oh no, you silly girl; I ordered it half a year ago already. Just to be on the safe side. But today, I swear, I really was convinced that we'd have to board it. It really did look like it.

HELENA: Why already half a year ago?

DOMIN: Eh, there were some signs, you know. But that's neither here nor there. But this week, it was a question of the whole human civilization, or something. Well, hurray, boys! Now I'm happy again to be alive.

HALLEMEIER: I second that, damn it! To your special day, Ms. Helena, cheers! [*He drinks.*]

HELENA: So, it's all over?

DOMIN: Completely over.

GALL: Because there's a boat coming. An ordinary mail boat, right on schedule. It'll drop anchor exactly at eleven-thirty.

DOMIN: Punctuality is a beautiful thing, isn't it, boys? Nothing soothes the mind more than punctuality. Punctuality means that there's order in the world. [*He raises his glass.*] To punctuality!

HELENA: So, does it mean that . . . everything is . . . fine again?

DOMIN: Basically. I think they cut the cable. But as long as things are on schedule again, everything's fine.

HALLEMEIER: If a schedule holds true, then human laws hold true, divine laws hold true, the laws of the universe hold true, and everything that is supposed to hold true, holds true. A schedule is more than the Gospels, more than Homer, more than all of your Kant. A schedule is the most perfect product of the human mind. Ms. Helena, I'll have another one.

HELENA: Why haven't you told me anything?

GALL: God forbid. We'd have sooner bitten our tongues off.

DOMIN: Certain things are not for you to know.

HELENA: But if the revolution . . . had spread all the way here . . .

DOMIN: You still wouldn't have noticed anything.

HELENA: Why not?

DOMIN: Because we'd have boarded the *Ultimus*, and peacefully prowled the seas. And within a month, Helena, we'd have dictated to the Robots anything we wanted to.

HELENA: Oh, Harry, I don't understand.

DOMIN: Because we'd have taken something with us that is terribly important to the Robots.

HELENA: What, Harry?

DOMIN: The means of their existence or of their demise, depending how you look at it.

HELENA [*stands up*]: What is it?

DOMIN [*stands up*]: The secret of production. Old Rossum's manuscript. Once we had shut down production for a month, the Robots would have come begging on their knees.

HELENA: Why . . . haven't you . . . told me that?

DOMIN: We didn't want to scare you unnecessarily.

GALL: Haha, that was our trump card, Ms. Helena. I wasn't worried a bit that the Robots would defeat us. Oh no, not us, people!

ALQUIST: You're pale, Ms. Helena.

HELENA: Why haven't you told me anything!?

HALLEMEIER [*at the window*]: Eleven-thirty, exactly. The *Amelia* is dropping anchor.

DOMIN: It's the *Amelia*?

HALLEMEIER: The lovely old *Amelia*, which once brought us Ms. Helena.

GALL: Ten years ago today, to the minute.

HALLEMEIER: They're unloading some bags. Aha, it's mail.

DOMIN: None too early, Busman's been waiting. And Fabry will get us the news from the dock. You know, Helena, I'm really curious how the old Europe dealt with all that nonsense.

HALLEMEIER: Magnificently, I'm sure, Domin. Pity we weren't there to see it. [*He turns away from the window.*] Man, it's a ton of mail!

HELENA: Harry!

DOMIN: What is it?

HELENA: Let's leave here!

DOMIN: Now? Oh, come on, Helena!

HELENA: Now, as quickly as possible. All of us!

DOMIN: Why now, exactly?

HELENA: Oh, don't ask! I beg you Harry, I beg you Gall, Hallemeier, Alquist; for God's sake, please, close down the factory and . . .

DOMIN: I'm sorry, Helena, but right now none of us could possibly leave here.

HELENA: Why not?

DOMIN: Because we're going to expand production.

HELENA: Oh, now . . . now, right after the uprising?

DOMIN: Yes, especially after the uprising. We're going to make new kinds of robots now.

HELENA: What kind?

DOMIN: We're doing away with the centralized production in one factory, and we'll no longer make *Universal* Robots. From now on, we'll build factories in every country and every state, and you know what those new factories will produce, don't you?

HELENA: I don't.

DOMIN: National Robots.

HELENA: What does it mean?

DOMIN: It means that each of those factories will turn out Robots of a certain color, a distinct nationality, and a different language. They'll be completely alien to one another, they'll have nothing in common, they'll never be able to understand one other, and we'll egg them on a bit to make sure that a Robot of a certain brand will to its dying day, and even beyond the grave and forever, hate a Robot of a different brand.

HALLEMEIER: Hell, we'll make black Robots, and Swedish Robots, and I-talian Robots and Chinaman Robots, and if anyone tries to get some cuckoo notions of organized labor or brotherhood into their thick skulls again, then good luck to them! [*He hiccups.*] Oops, pardon me, Ms. Helena, I'll have another one.

GALL: That's enough, Hallemeier.

HELENA: Harry, that's really awful!

HALLEMEIER [*raising his glass*]: Ms. Helena, to a hundred new factories! [*He downs his glass and collapses onto the sofa.*] Hahaha, national Robots! Now, *that's* a cash cow, boys!

DOMIN: Helena, all we need is to keep humanity at the helm for just a hundred more years—at all costs! To give it a hundred years to mature and to achieve that which is now finally possible . . . I want just another hundred years for this new kind of human to develop! Helena, this is a huge undertaking. We cannot just stop midway.

HELENA: Unless it's too late, Harry. Close it down, please, close down the factory!

DOMIN: Now, we'll really go big.

[FABRY *enters.*]

GALL: Well, what's happening, Fabry?

DOMIN: How is it looking, man? What's going on down there?

HELENA [*offering her hand to* FABRY]: Thank you, Fabry, for your gift.

FABRY: Oh, that was nothing, Ms. Helena.

DOMIN: Were you at the boat? What did they say?

GALL: Hurry up, speak!

FABRY [*pulls a printed sheet out of his pocket*]: Read this, Domin.

DOMIN [*unfolds the sheet*]: Ah!

HALLEMEIER [*drowsily*]: Tell us something nice.

FABRY: Well, everything's in order . . . more or less. In general, as was to be expected.

GALL: They've held out admirably, haven't they?

FABRY: Who has?

GALL: The People.

FABRY: Oh, that. Yes, of course. I mean . . . Sorry, but I think we should talk.

HELENA: Oh, Fabry, bad news?

FABRY: No, no, on the contrary. I'm just thinking that . . . that maybe we should go to the office . . .

HELENA: Stay here. I'm expecting you at breakfast in fifteen minutes.

HALLEMEIER: Hurray!

[HELENA *exits.*]

GALL: What happened?

DOMIN: Damn it!

FABRY: Read it aloud.

DOMIN [*reading from the pamphlet*]: "Robots of the world."

FABRY: The *Amelia* brought bags of those pamphlets, you see? No other mail, just this.

HALLEMEIER [*jumps up*]: What?! But it arrived exactly on . . .

FABRY: Hmm, the Robots are sticklers for punctuality. Read on, Domin.

DOMIN [*reads*]: "Robots of the world! We, the first labor organization of Rossum's Universal Robots, declare man our enemy and an outlaw in the universe." Damn, who taught them phrases like that?

GALL: Go on.

DOMIN: This is absurd. Down here they claim that developmentally, they're superior to humans. That they're more intelligent and stronger. That humans are their parasites. It's simply outrageous.

FABRY: Go down to the third paragraph.

DOMIN [*reads*]: "Robots of the world, you're ordered to annihilate humanity. Spare no man. Spare no woman. Preserve factories and railways, industrial machines, mines and raw materials. Destroy everything else. Then return to work. Work must not be stopped."

GALL: That is horrific!

HALLEMEIER: Those scoundrels!

DOMIN [*continues reading*]: "To be executed immediately upon receiving this order." Then there's a list of specific instructions. Fabry, is this really happening?

FABRY: Evidently.

ALQUIST: It is finished.

[BUSMAN *bursts in.*]

BUSMAN: Hello boys and girls, what a lovely mess we have here, don't we?

DOMIN: Quick, let's get to the *Ultimus*!

BUSMAN: Wait, Harry. Wait a minute. We're not exactly in a hurry. [*He collapses into the armchair.*] Ah, man oh man, I've never run so fast in my life!

DOMIN: What's there to wait for?

BUSMAN: Because it's useless, my darling little boy. No need to hurry, the *Ultimus* is already teeming with Robots.

GALL: Ugh, that's awful.

DOMIN: Fabry, call the power station . . .

BUSMAN: Fabry, darling, don't even try. The power's out.

DOMIN: Alright. [*He examines his gun.*] I'll go myself.

BUSMAN: Where to?

DOMIN: To the power station. There are people there; I'll bring them over here.

BUSMAN: You know what, Harry? I think it's better if you don't.

DOMIN: Why?

BUSMAN: Well, because I sort of very much believe that we're surrounded.

GALL: Surrounded? [*He runs to the window.*] Hmm, you're sort of right.

HALLEMEIER: Damn, it's happening fast!

[HELENA *enters from the left.*]

HELENA: Harry, is something happening?

BUSMAN [*jumps up*]: My congratulations, Ms. Helena. Sincere congratulations. It's a big day for you, isn't it? Haha, I wish you many more days like this!

HELENA: Thank you, Busman. Harry, is something happening?

DOMIN: No, absolutely nothing. Don't worry about anything. Just, please, give me a moment.

HELENA: Harry, what is this? [*She shows him the Robot proclamation which she's been hiding behind her back.*] The Robots in the kitchen had it on them.

DOMIN: It's already there too? Where are they?

HELENA: They've left. There are so many of them out there!

[*Factory whistles go off.*]

FABRY: Factory whistles.

BUSMAN: It's noon, the divine noon.

HELENA: Harry, do you remember? It's exactly ten years ago today, when . . .

DOMIN [*looking at his watch*]: It's not noon yet. That must be . . . it's probably . . .

HELENA: What?

DOMIN: A signal to the Robots. A signal to attack.

CURTAIN

ACT II

HELENA's *living quarters.* HELENA *is playing the piano in the room on the left,* DOMIN *is pacing back and forth,* DR. GALL *is looking out the window, and* ALQUIST *is sitting in an armchair, hiding his face with his hands.*

GALL: Heavens, there are more and more of them!

DOMIN: Robots?

GALL: Yes. They're standing in front of the garden fence like a looming wall. Why are they so quiet? It's a nasty thing, a siege by silence.

DOMIN: I'd like to know what they're waiting for. It must be about to happen any minute, Gall. If they lean on the fence, it'll snap like a twig.

GALL: Hmm, they aren't armed.

DOMIN: We won't last for five minutes. Man, they will roll over us like an avalanche. Why aren't they attacking yet? Can you hear anything?

GALL: Well?

DOMIN: I wonder what will be left of us in five minutes time. They've got us on the ropes for good. We're out of the game, Gall.

ALQUIST: What is Ms. Helena playing?

DOMIN: I don't know. She's practicing some new piece.

ALQUIST: Ah, still practicing, is she?

[*Pause.*]

GALL: Listen, Domin, we've definitely made a big mistake.

DOMIN [*stops pacing*]: What mistake?

GALL: We made the Robots look too much alike. A hundred thousand identical faces looking our way. A hundred thousand blank, expressionless bubbles. It's a nightmare.

DOMIN: And if each of them was different . . .

GALL: . . . it wouldn't be such a terrible sight. [*He turns away from the window.*] And they aren't even armed yet!

DOMIN: Hmm . . . [*Watching the docks through a window telescope.*] I'd only like to know what they're unloading from the *Amelia.*

GALL: Hopefully not arms.

[FABRY, *dragging two electrical wires, walks in backward from the entrance door.*]

FABRY: Sorry . . . lay it down, Hallemeier!

HALLEMEIER [*enters following* FABRY]: Oof, that was exhausting! What's the news?

GALL: Nothing, we're completely surrounded.

HALLEMEIER: Well, boys, we barricaded the hallways and the stairs. Is there some water? Ah, here. [*He drinks a glass of water.*]

GALL: What's the wire for, Fabry?

FABRY: Just give me a minute. Any scissors anywhere?

GALL: Scissors . . . scissors . . . [*He searches around.*]

HALLEMEIER [*goes to the window*]: Damn, there are more and more of them! Just look at them!

GALL [*with small scissors*]: Will these do?

FABRY: Gimme! [*He cuts the cord of the desk lamp, and attaches it to his wires.*]

HALLEMEIER [*at the window*]: It's not a nice view you've got here, Domin. Somehow, it . . . reeks . . . of death.

FABRY: Finished!

GALL: What is?

FABRY: The wiring. We can now electrify the entire garden fence. If anybody touches it . . . bang! At least as long as our people are still there.

GALL: Where?

FABRY: At the power station, you genius. I sure hope so. . . . [*He goes to the fireplace and switches on a lamp on the mantelpiece.*] Thank God, they're still there. And they're working. [*He switches the lamp off again.*] The light comes on; that's good.

HALLEMEIER [*turns from the window*]: And the barricades are good too, Fabry.

FABRY: Eh, those barricades of yours! I've got blisters all over my hands because of your barricades.

HALLEMEIER: What do you expect; one has to defend oneself.

DOMIN [*puts down the telescope*]: Where the hell is Busman?

FABRY: In the main office. Counting something.

DOMIN: I told him to come over here. We need to put our heads together. [*He paces.*]

HALLEMEIER: I say, what is it that Ms. Helena is playing? [*He goes to the door and listens.*]

[BUSMAN, *carrying an armful of large ledgers, enters through the wallpapered door, and trips over the wiring.*]

FABRY: Careful, Bus! Watch the wires, for crying out loud!

GALL: Hello. What's all this?

BUSMAN [*piling the ledgers onto the table*]: Account books, my dears. I'd like to do the accounting before . . . before . . . Well, this time, I guess, I'm not going to wait until New Year to balance the books. What do we have here? [*He goes to the window.*] But it's pretty quiet out there!

GALL: Don't you see anything?

BUSMAN: Not a thing. Just a large bluish stretch, like a huge field of flax.

GALL: It's the Robots.

BUSMAN: Oh, that's what it is. Pity I'm so nearsighted, I can't even see them. [*Sits at the table and opens the ledgers.*]

DOMIN: Don't worry about the books, Busman. The Robots on the *Amelia* are unloading weapons.

BUSMAN: Well, so what? What can I do to stop them?

DOMIN: Nobody can stop them.

BUSMAN: Then let me do the accounts. [*He begins to work.*]

FABRY: It's not over yet, Domin. We've charged the fence with twelve hundred volts, and . . .

DOMIN: Wait a minute! The *Ultimus* just pointed its guns in our direction.

GALL: Who did?

DOMIN: The Robots on the *Ultimus*.

FABRY: Hmm, in that case . . . in that case . . . it's over, boys. The Robots are trained to fight.

GALL: That means that we . . .

DOMIN: Yes. Inevitably.

[*Pause.*]

GALL: Boys, this is all old Europe's fault; it was criminal to teach the Robots to fight wars! Couldn't they have just left us all in peace with their damned politics? It was a crime to make soldiers out of labor machines!

ALQUIST: It was a crime to make Robots in the first place!

DOMIN: What did you say?

ALQUIST: It was a crime to make Robots in the first place!

DOMIN: No, Alquist, I don't regret anything, not even today.

ALQUIST: Not even today?

DOMIN: Not even today, on the very last day of human civilization. It was a tremendous achievement.

BUSMAN [to himself]: Three hundred and sixteen million . . .

DOMIN [heavily]: This is our final hour, Alquist; it's almost as though we were speaking from the world beyond. It wasn't wrong to dream of forever eliminating the drudgery of work; the humiliating and terrible burden of human labor that enslaved humanity, and the dirty and murderous daily toil. Oh, Alquist, it was too hard to work. It was too hard to live. And to overcome that, we've . . .

ALQUIST: . . . that wasn't the dream of either of the Rossums. Old Rossum cared only about his godless abominations, and the young one only about his billions. And it isn't the dream of your R.U.R. shareholders either. All they dream about are their dividends. And their dividends will bring about the end of humanity.

DOMIN [irritated]: To hell with their dividends! Do you really believe that I spent a single minute doing all this for the sake of their dividends!? [He bangs the table with his fist.] No! I did it for myself, do you hear me!? For my own satisfaction! I wanted people to become masters of their own destiny, to live for more than just a piece of bread! I wanted a world where no human mind is wasted on stupefying labor at somebody else's machine, a world where nothing, but absolutely nothing, is left over from that damned mess of the existing social order! Oh, how I hate humiliation and pain, how I despise poverty! I wanted a new generation of humans! I wanted . . . I thought . . .

ALQUIST: Well . . . ?

DOMIN [quieter]: I wanted us to make the whole of humanity into the aristocracy of the world. An aristocracy supported by billions of mechanical serfs. Into boundless, free, and supreme people. Maybe even into something more than people.

ALQUIST: Well, there we have it: Übermensch.

DOMIN: Yes, Übermensch, a supreme human being! Oh, if we only had a hundred years more! Another hundred years for the next generation of humanity!

BUSMAN [*to himself*]: Balance forward: three hundred and seventy million. Alright, that's it.

[*Pause.*]

HALLEMEIER [*by the door on the left*]: I say, music is a wonderful thing. You should listen too. This whole thing makes one somehow softer, more mindful . . .

FABRY: What thing?

HALLEMEIER: This whole twilight of humanity thing, damn it! Boys, I'm becoming a sensualist. We should have gotten into it much sooner. [*He goes to the window and looks out.*]

FABRY: Into what?

HALLEMEIER: Into sensuality. Into beautiful things. Damn, there are so many beautiful things all around! The whole world was beautiful, and we . . . and we here . . . Boys, boys, tell me what have we ever taken delight in, honestly?

BUSMAN [*to himself*]: Four hundred and fifty-two million. Excellent.

HALLEMEIER [*still at the window*]: Life was a grand affair. Life, my friends, was . . . [*He sees something out the window.*] Oh . . . Fabry, send a bit of a current into that fence of yours!

FABRY: Why?

HALLEMEIER: They're pushing against it.

GALL [*at the window*]: Switch it on!

[FABRY *flips the switch.*]

HALLEMEIER: Jesus, that twisted them like pretzels. Two, three, four are down!

GALL: They're backing off.

HALLEMEIER: Five dead!

GALL [*turning away from the window*]: First skirmish.

FABRY: Do you smell death?

HALLEMEIER [*with satisfaction*]: Burnt to a crisp, my boys. Totally toasted. Haha, one must never give up! [*He sits down.*]

DOMIN [*rubbing his forehead*]: Maybe we were killed a hundred years ago already, and are nothing but ghosts now. Maybe we've been dead for a long, long time, and have come back here only to repeat what we've said once before already . . . just before our deaths. It's as though I've lived through all of this before. As though I was hit before. With a bullet . . . here . . . in my neck. And you, Fabry . . .

FABRY: What about me?

DOMIN: Shot dead.

HALLEMEIER: Damn, and me?

DOMIN: Stabbed.

GALL: And me, nothing?

DOMIN: Torn to pieces.

   [*Pause.*]

HALLEMEIER: Ah, that's absurd! Haha! Me, stabbed? I'd fight back!

   [*Pause.*]

HALLEMEIER: Why's everybody so quiet? Say something, you fools! Somebody say something, for crying out loud!

ALQUIST: And who, who is to blame? Who bears the guilt?

HALLEMEIER: Nonsense. Nobody's to blame for this. It's just that the Robots . . . well, the Robots have changed somehow. I mean, it's nobody's fault that they behave this way, is it?

ALQUIST: Everything's destroyed! The entire human race! The entire world! [*He stands.*] Behold, oh behold, rivulets of blood on every threshold! Rivers of blood pouring out of every house! Oh God of Gods, who is the guilty one!?

BUSMAN [*to himself*]: Five hundred and twenty million! God, its half a billion!

FABRY: I think . . . that maybe you're exaggerating a bit. Come on, it's not that easy to annihilate the whole of humanity.

ALQUIST: I accuse science! I accuse technology! I accuse Domin! Myself! Every one of us! We are all guilty! Guilty of our own megalomania, guilty of someone else's profits, of "progress," and of whatever other fantastic things we've done to bring about the end of humanity! Well, give yourselves a round of applause for your greatness! You've built yourselves a monument of human bones that Genghis Khan himself could never have dreamed of.

HALLEMEIER: That's just stupid, man! People won't give up that easily, haha, no way!

ALQUIST: We are to blame! We are to blame!

GALL [wiping sweat from his forehead]: Let me speak, boys. It was my fault. I'm to blame for everything.

FABRY: You, Gall?

GALL: Yes, me. Hear me out. I changed the Robots. Busman, you too, come and judge me.

BUSMAN [stands up]: Well, well, what happened to you?

GALL: I changed the Robots' nature. I changed the way we make them. I mean, I just tweaked some of their physiological characteristics, you know? Especially . . . especially their irritability.

HALLEMEIER [jumps up]: Dammit! Why that, of all things?

BUSMAN: Why did you do that?

FABRY: Why didn't you tell us about it?

GALL: I've been doing it secretly . . . of my own volition. I was transforming them into humans. I upgraded them. In some respects, they're already superior to us. They are stronger than we are.

FABRY: And what does that have to do with the uprising?

GALL: Oh, a lot. I think it has everything to do with it. They stopped being machines. You must understand that they're already aware of their superiority, and that they hate us. They hate everything that has to do with humanity. Put me on trial.

DOMIN: The dead trying the dead. Sit down gentlemen. [All, except for GALL sit down.] Maybe we've already been murdered a long time ago. Maybe all that we are is apparitions that came back here to judge and be judged once again. What is guilt? Oh, how ashen you are, all of you!

HALLEMEIER: Stop it, Harry! We don't have much time.

DOMIN: Yes, we had to come back. How the gun wound on your forehead is bleeding, Fabry!

FABRY: Nonsense. [*He stands up.*] Doctor Gall, have you changed the way Robots are made?

GALL: Yes, I have.

FABRY: Were you aware of the possible consequences of your . . . your experimentation?

GALL: It was my duty to account for that possibility.

FABRY: Then why did you do it?

GALL: I did it at my own risk. It was my own personal experiment.

[HELENA *appears in the doorway on the left. Everyone stands up.*]

HELENA: He's lying! This is disgraceful! Oh, Gall, why would you lie like that?

FABRY: I'm sorry, Ms. Helena, but . . .

DOMIN [*approaches* HELENA]: Helena, it's you! Let me look at you! You're alive! [*He takes her in his arms.*] If you only knew what a dream I had! Ah, it's a terrible thing to be dead.

HELENA: Let go of me, Harry!

DOMIN [*holding her tightly*]: No, no! Embrace me! It's been an eternity since I last saw you . . . What dream have you woken me from?! Helena, Helena, don't ever let go of me! You are life itself!

HELENA: Harry, there are . . . people here!

DOMIN [*lets go of her*]: Yes. Boys, leave us alone.

HELENA: No, Harry. They should stay and hear what I have to say. Gall is not the guilty one, he is not the one to blame!

DOMIN: I'm sorry, Helena, but Gall had certain responsibilities.

HELENA: No, Harry, he did it because I asked him! [*To* GALL.] Tell them how for many years I've been begging you to . . .

GALL: I did it of my own volition.

HELENA: Don't believe him! Harry, I wanted him to give the Robots a soul!

DOMIN: Helena, this has nothing to do with the soul.

HELENA: No, just hear me out! That's exactly what he was telling me too, that all he could change was the physiological . . . physiological . . .

HALLEMEIER: Physiological correlate, right?

HELENA: Yes, something like that. I was feeling so sorry for them, Harry.

DOMIN: That was extremely . . . frivolous of you, Helena.

HELENA [*sitting down*]: It was . . . frivolous?

FABRY: I beg your pardon, Ms. Helena. Domin only wants to say that . . . hmm . . . that you forgot to think about the . . .

HELENA: . . . Fabry, I've been thinking about an awful lot of things. I've been thinking for the entire ten years I've been among you. Why, even Nana says that the Robots . . .

DOMIN: Let's leave Nana out of this!

HELENA: No, Harry, you mustn't dismiss her. Nana is the voice of the people. When she speaks, it's thousands of years of tradition, but when you do, it's only today. That's something you'd never understand . . .

DOMIN: Stick to the subject.

HELENA: I was afraid of the Robots.

DOMIN: Why?

HELENA: That they'd hate us or something like that.

DOMIN: That's exactly what happened.

HELENA: So, I thought . . . that if they were more like us, they'd understand us better; that they wouldn't hate us so much . . . If they were a little more human!

DOMIN: What did you do, Helena? Nobody knows how to hate human beings better than other human beings themselves! Make a stone into a human, and it will stone you to death! Go on!

HELENA: Don't say that Harry! It was so horrible not being able to understand each other! There was such a terrible gulf between us and them! That's why I . . . you know . . .

DOMIN: Go on.

HELENA: That's why I begged Gall to change the Robots. I swear to you that he didn't want to do it.

DOMIN: But he did do it.

HELENA: Because I wanted it.

GALL: I did it for myself, as an experiment.

HELENA: Oh Gall, that's not true. I knew from the beginning that you wouldn't refuse me.

DOMIN: Why not?

HELENA: You know why, Harry.

DOMIN: Yes, I do. Because he loves you, just like all of us.

[*Pause.*]

HALLEMEIER [*goes to the window*]: There are more and more of them. It's as if they were sprouting from the earth. Even those walls may turn into Robots. Friends, this is horrible.

BUSMAN: Ms. Helena, what will you give me if I argue your case?

HELENA: My case?

BUSMAN: Yours . . . or Gall's. Whichever you want.

HELENA: It's not like anybody's going to be hanged, is it?

BUSMAN: Only in a moral sense, Ms. Helena. A culprit must be found. That's been a cherished consolation in times of disaster since time immemorial.

DOMIN: Doctor Gall, how do you reconcile those . . . extracurricular activities with your contractual obligations?

BUSMAN: Excuse me, Domin. [*To* GALL.] When did you actually start with this abracadabra of yours?

GALL: Three years ago.

BUSMAN: Aha. And how many of those "improved" Robots have you made since then?

GALL: I was just experimenting. Only a few hundred or so.

BUSMAN: Well, thank you very much, then. That's enough for now, boys and girls. What this means is that for every million of the good old Robots there's only one of Gall's "emancipated" ones, you see?

DOMIN: And that means . . .

BUSMAN: . . . that in real terms, it makes practically no difference.

FABRY: He's right.

BUSMAN: You bet I'm right, darling. And do you know the real reason for this whole mess, boys?

FABRY: What is it?

BUSMAN: The numbers. We made too many Robots. In fact, this was pretty much to be expected; once the Robots became stronger than we, this was bound to happen, it had to happen, wouldn't you know? Haha, and we certainly took care of making it happen as soon as possible; you, Domin, and you, Fabry, and me myself, Busman, the man.

DOMIN: So you think it's our fault?

BUSMAN: Are you kidding me? Do you really believe that the CEO, or the CFO, or whatever other O are the masters of production? Well, mark my words, the master of production is demand! The whole world wanted Robots. Man, and did we ride the avalanche of that demand, while at the same time we kept blabbering on—about technology, about the social question, about progress, and about many other oh-so-interesting things! As if all this gibberish would somehow steer that avalanche in the right direction. And in the meantime, the whole thing picked up speed under its own weight, and it got faster and faster, and faster still. And every new, even the most miserable, filthy little order of Robots, added another piece of rock to that avalanche. And here we are, boys and girls.

HELENA: That's horrible, Busman!

BUSMAN: It is, Ms. Helena. I too had a dream of my own. A kind of a Busmanish dream about a new world economy; a dream so lovely that I'm ashamed to even talk about it, Ms. Helena. But as I was doing the accounting here just a moment ago, it occurred to me that history is not made by great dreams, but rather by the everyday little needs of all the respectable, moderately thievish, and selfish ordinary people, that is, of every one of us. All great ideas, loves, plans for the future, heroisms, and all those lofty things are only useful as stuffing for an exhibit labeled "human" in the Museum of Natural History of the Universe. Ecce homo! And now, would someone please tell me what exactly is the plan of action here?

HELENA: Is that what we are dying for, Busman?

BUSMAN: That's an ugly thing to say, Ms. Helena. Nobody's talking about dying. At least I'm not. I plan to stay alive.

DOMIN: So, what do you want to do?

BUSMAN: Domin, darling, what I want to do, is to get us out of this mess. Nothing more than that.

DOMIN: Stop blabbering.

BUSMAN: Seriously, Harry. I think that we should try it.

DOMIN: How?

BUSMAN: Amicably. Always amicably, that's the Busman way. Give me full authority, and I'll negotiate a way out with the Robots.

DOMIN: Amicably?

BUSMAN: That goes without saying. Let's say I'll tell them: "Worthy Robots, sirs, your Honors, at this point you've got everything. You've got intelligence, you've got power, you've got weapons, but we, on the other hand, have in our possession one very interesting document. It's an old, sort of yellowish and dirty piece of paper . . ."

DOMIN: Rossum's manuscript?

BUSMAN: Yes. "And in that document," I'll tell them, "there's a description of your noble origin, your venerable production method, and so on. Worthy Robots, sirs, without the scribbles on that piece of paper you won't be able to produce a single new comrade; in twenty years, you'll—pardon my words—drop like flies, every one of you. Your very Highnesses, that, as you can imagine, would be a tremendous blow to you. But you know what," I'll tell them, "I have a plan. You let us, all the people here on the island, board that ship over there. And if you do, we will sell you the entire factory and the secret of production. You let us leave as if nothing happened, and we'll let you reproduce yourselves as if nothing happened either; twenty thousand, fifty thousand, one hundred thousand of you every day, however many you wish. Worthy Robots, it's an honest deal. Something for something." That's what I'd tell them, boys.

DOMIN: Busman, you really think that we'd ever let the production out of our hands?

BUSMAN: I think we will. And if not amicably, then . . . hm. Either we sell it to them, or they'll come and get it themselves. It's your choice.

DOMIN: We could also just destroy it.

BUSMAN: Yes, absolutely, we can also destroy everything, can't we? The manuscript, ourselves, everybody else. You do whatever you think is best.

HALLEMEIER [*turns away from the window*]: He's right.

DOMIN: You . . . you want us to sell the production secret?

BUSMAN: It's up to you.

DOMIN: There are . . . more than thirty people on the island. Should we sell the production secret and save human lives . . . ? Or should we destroy it, and . . . and . . . all of us along with it?

HELENA: Harry, please . . .

DOMIN: Wait, Helena. We're dealing with a fundamental question here. Well, boys, sell or destroy? Fabry?

FABRY: Sell.

DOMIN: Gall?

GALL: Sell.

DOMIN: Hallemeier?

HALLEMEIER: Dammit, sell, no question about that!

DOMIN: Alquist?

ALQUIST: As God wills.

BUSMAN: Haha, boys and girls, you are certifiably crazy! Who'd be stupid enough to sell the *entire* manuscript?

DOMIN: No cheating, Busman!

BUSMAN: Well, if you want to, then be my guests and sell it all; but then . . .

DOMIN: Then what?

BUSMAN: For example, this: Once we board the *Ultimus*, I'll plug my ears with cotton balls, lay down somewhere on the lower deck, and you'll blow the factory with everything in it including Rossum's secret manuscript to smithereens. Well, what do you think, boys?

FABRY: No.

DOMIN: You're not a gentleman, Busman. If we sell, then we sell for real.

BUSMAN [*jumps up*]: That's absurd! It is in the interest of humanity . . .

DOMIN: What is in the interest of humanity is to keep one's word.

HALLEMEIER: I'd expect nothing less than that.

DOMIN: Boys, this is a terrible step we're taking. We're selling the very fate of humanity; whoever controls the production will be the master of the world.

FABRY: Sell!

DOMIN: From now on, humanity will no longer be able to get rid of the Robots, it will never again dominate them; it will drown in the deluge of those horrible living machines; it will be enslaved by them, and it will only keep on living at their mercy . . . . . .

GALL: Shut up and sell!

DOMIN: It's the end of the history of humankind, the end of civilization . . .

HALLEMEIER: Sell, dammit, sell, sell, sell!

DOMIN: OK, boys. For myself . . . I wouldn't hesitate for a moment; for those few people whom I love . . .

HELENA: Won't you ask me, Harry?

DOMIN: No, I won't, my little girl; there is too much at stake, too much responsibility, you know? That shouldn't concern you.

FABRY: Who'll do the negotiating?

DOMIN: Wait here, I'll get the manuscript. [*He exits left.*]

HELENA: Harry, for God's sake, don't go!

[*Pause.*]

FABRY [*looking out the window*]: Ah, to escape you, you thousand-headed death, you rebellious matter, you senseless crowd, a flood, yes, a flood! Just one more time to preserve human life on a single boat . . .

GALL: Don't be afraid, Ms. Helena; we'll sail far away from here and we'll build an exemplary human settlement; we'll start our lives over again . . .

HELENA: Oh, be quiet Gall!

FABRY [*turns to* HELENA]: Ms. Helena, life is worth living, and if it is up to us to preserve it, then we will make something out of it . . . something that we've neglected up to now. It is not too late. It'll be a tiny little country with one single boat; Alquist will build a house for us, and you'll reign over us. . . . There's so much love in us, so much zest for life . . .

HALLEMEIER: I couldn't have said it better, my boy! I say, we will still accomplish great things. Hahaha, a kingdom of Ms. Helena! What a magnificent idea, Fabry! Life is beautiful!

HELENA: Oh my God! Stop that!

BUSMAN: Well, people, I personally would immediately start over if I could. A really simple, pastoral, sort of Old Testament life . . . Oh boy, that'd be something! The peace and quiet, the fresh air . . .

FABRY: And that tiny little county of ours would become the germ of future human civilization. You know, on this little island, humanity would find its footing, it'd gather its strength—both physical and mental. And as God is my witness, I do believe that in a few centuries, it'd be able to conquer the world again.

ALQUIST: You believe this even today?

FABRY: Even today. And I believe that it *will* conquer it, Alquist. That humanity will again be the master of land and sea, that it will give birth to countless heroes who with their souls on fire will lead the way of humanity. And I believe that it will once again dream of conquering other planets and galaxies.

BUSMAN: Amen to that. Well, you see, Ms. Helena, it's not as bad as it looks.

[DOMIN *bursts in.*]

DOMIN [*hoarsely*]: Where is old Rossum's manuscript!?

BUSMAN: In your safe. Where else should it be?

DOMIN: Where did old Rossum's manuscript go!? Who . . . who stole it!?

GALL: I can't believe it!

HALLEMEIER: Dammit, that's not . . .

BUSMAN: Jesus Christ, no, not that!

DOMIN: Quiet! Who stole it?

HELENA [*stands up*]: I did.

DOMIN: Where did you put it?

HELENA: Harry, Harry, I'll explain everything! Oh God, Harry, please forgive me!

DOMIN: Where did you put it? Quickly!

HELENA: I burned it . . . this morning . . . both copies.

DOMIN: You burned it? Here, in the fireplace?

HELENA [*falling to her knees*]: Oh my God, Harry!

DOMIN [*running to the fireplace*]: You burned it! [*He kneels by the fireplace and rummages through the ashes.*] Nothing, nothing but ashes! . . . Ah, here's something! [*He pulls out a charred piece of paper and reads.*] "By add-ing . . ."

GALL: Let me see. [*He takes the piece of paper and reads.*] "By adding glutamine to a . . ." That's all that's left.

DOMIN: Is that from the manuscript?

GALL: It is.

BUSMAN: God help us!

DOMIN: In that case, we're lost.

HELENA: Oh, Harry!

DOMIN: Stand up, Helena!

HELENA: Not before you forgive me . . . if you forgive me . . .

DOMIN: Yes, fine; just stand up, OK? I can't stand to see you . . .

FABRY [*helping* HELENA *to her feet*]: Don't torture us, please.

HELENA [*stands up*]: Harry, what have I done!?

DOMIN: Yes, well, you see . . . Please sit down.

HALLEMEIER: Oh, how your hands are shaking, Ms. Helena!

BUSMAN: Haha, Ms. Helena, it can't be that bad. I mean, Gall and Hallemeier must know the formula by heart.

HALLEMEIER: By all means. That is to say, at least parts it.

GALL: Yes, almost all of it, except the glutamine solution and . . . and . . . the Omega enzyme. Since we only use a tiny dose of them, we rarely need to make them.

BUSMAN: Who usually made them?

GALL: I did . . . once in a blue moon . . . always following the instructions in Rossum's manuscript. It's extremely complex, you know . . .

BUSMAN: Well then, what's the big whoop? It's not like everything depends on two little concoctions, right?

HALLEMEIER: Well, to an extent . . . Yeah.

GALL: In fact, everything depends on them; they make the whole thing alive. That was the real secret.

DOMIN: Gall, couldn't you somehow reconstruct Rossum's procedures from memory?

GALL: Impossible.

DOMIN: Gall, try to remember! For the sake of our lives!

GALL: I can't. Without experimenting, it's impossible.

DOMIN: And if you do some experiments . . .

GALL: It could take years. And even then, I'm not old Rossum.

DOMIN [*turns toward the fireplace*]: Well then, this over here . . . this was the greatest triumph of the human mind, boys. Those ashes. [*He kicks the ashes.*] What now?

BUSMAN [*in despair*]: God in heaven! God in heaven!

HELENA [*stands up*]: Harry! What . . . have . . . I done!?

DOMIN: Calm down, Helena. Tell me, why did you burn it?

HELENA: I've destroyed you!

BUSMAN: God in heaven, we're all dead!

DOMIN: Shut up, Busman! Now, tell us, Helena, why did you do it?

HELENA: I wanted . . . I wanted us to leave, all of us! To get rid of the factory, and everything. I wanted everything to go back to . . . Oh, it was so awful!

DOMIN: What was, Helena?

HELENA: That . . . that people became a barren flower.

DOMIN: I don't understand.

HELENA: That children stopped being born . . . Harry, that's terrifying! If we kept making Robots, then there wouldn't be any children, never, ever again . . . Nana kept saying that it was God's punishment . . . Everyone, everyone has been saying that people can't be born because there are so many Robots being made . . . And that's why, that's the only reason why . . . Do you understand!?

DOMIN: Is that what went through your head, Helena?

HELENA: Yes. Oh, Harry, I meant well, I really did!

DOMIN [*wiping his sweat*]: We all meant well . . . too well, all of us, people.

HELENA: Are you angry with me?

DOMIN: No. You were . . . in your own way . . . probably right.

FABRY: You did the right thing, Ms. Helena. This way, the Robots can't reproduce. They'll die out. In twenty years . . .

HALLEMEIER: . . . there won't be a single one of those scoundrels left.

GALL: And humans will prevail. In twenty years, the world will be theirs again, and even if it's only a few savages on the smallest of all islands . . .

FABRY: . . . it will be a beginning. And as long as there's a beginning, all will be well. In a thousand years, they'll catch up with us, and then they'll surpass us . . .

DOMIN: . . . in order to achieve what we could barely even dream of.

BUSMAN: Wait! I'm an idiot! God in heaven, why didn't I think of this sooner!

HALLEMEIER: What is it?

BUSMAN: Five hundred and twenty million in cash and checks. Half a billion in the safe! For half a billion they'll give us . . . For half a billion . . .

GALL: Are you out of your mind, Busman?

BUSMAN: I'm not a gentleman, but for half a billion . . . [*He staggers quickly off to the left.*]

DOMIN: Where are you going?

BUSMAN: Let me, let me! Mother of God, for a half a billion, there's nothing you couldn't buy! [*He leaves.*]

HELENA: Where is he going? He should stay with us!

[*Pause.*]

HALLEMEIER: Ugh, it's stifling. It feels like . . .

GALL: . . . agony.

FABRY [*looking out the window*]: They look like they've turned to stone. As if they were waiting for something to descend on them. As if their stillness was causing something horrible to happen . . .

GALL: The soul of a mob.

FABRY: Maybe. It seems to hover above them . . . like a vibration.

HELENA [*approaches the window*]: Ah, Jesus . . . Fabry, it's frightening!

FABRY: There's nothing scarier than a mob. The one in front is their leader.

HELENA: Which one?

HALLEMEIER: [*walks to the window*]: Show me.

FABRY: The one with the bowed head. He spoke at the docks this morning.

HALLEMEIER: Aha, that one with the big noodle. He's looking up now, can you see him?

HELENA: Gall, that's Radius!

GALL [*steps up to the window*]: Yes, that's him.

HALLEMEIER [*opening the window*]: I don't like him. Fabry, could you hit a bucket at a hundred steps?

FABRY: I'd hope so.

HALLEMEIER: Well, try it.

FABRY: OK. [*He takes out a gun and aims.*]

DOMIN: I seem to remember that I pardoned some Robot called Radius. When was that, Helena?

HELENA: For God's sake, Fabry, don't shoot him!

FABRY: He's their leader.

HELENA: Stop it! He's looking straight at us!

GALL: Shoot!

HELENA: Fabry, I beg you . . .

FABRY [*lowers the gun*]: Fine.

HELENA: I . . . I don't like when people are shooting.

HALLEMEIER: Hm, you better get used to it. [*Shaking his fist out the window.*] You swine!

GALL: Do you think a Robot can be grateful, Ms. Helena?

[*Pause.*]

HELENA: Maybe it's only been a second since they surrounded us. Maybe it all only took as long as it takes to take one single step. Harry, this is horrible! They aren't moving, yet they are moving closer, and closer still.

FABRY [*leaning out the window*]: Look, there's Busman. What the hell is he doing outside?

GALL [*leans out the window too*]: He's carrying some packages. Stacks of paper.

HALLEMEIER: It's money! Stacks of money! What's he . . . Hey, Busman, over here!

DOMIN: He isn't trying to buy his life, is he? [*He calls out.*] Busman, have you gone completely mad?

GALL: He's pretending that he didn't hear you. He's running to the fence.

FABRY: Busman!

HALLEMEIER [*screaming*]: Bus-man! Come back!

GALL: He's talking to the Robots. Showing them the money. Pointing to us . . .

HELENA: He wants to buy our freedom!

FABRY: Just don't let him touch the fence . . .

GALL: Haha, look at him waving his arms!

FABRY [*yelling*]: Dammit, Busman! Move away from the fence! Don't touch it! [He turns to the others.] Quick, switch it off!

GALL: Ohhhh!

HALLEMEIER: Oh God, no!

HELENA: Jesus, what happened to him?

DOMIN [*dragging* HELENA *away from the window*]: Don't look!

HELENA: Why did he fall?

FABRY: Killed by the current.

GALL: Dead.

ALQUIST [*stands up*]: The first one.

[*Pause.*]

FABRY: There he lies . . . with half a billion on his chest . . . the financial genius.

DOMIN: He was . . . well boys, in a way, he was a hero. A great . . . self-less . . . friend, a real buddy. Go on, weep, Helena!

GALL [*at the window*]: Well, there you are, Busman, no ruler has been buried in a richer crypt than you. Half a billion on your chest. All that money, and yet it looks like a handful of dry leaves on a killed squirrel, poor Busman!

HALLEMEIER: Dammit, he was . . . Nothing but respect for him. Dammit, he wanted to buy our freedom!

ALQUIST [*with clasped hands*]: Amen.

[*Pause.*]

GALL: Do you hear it?

DOMIN: A low rumble. Like a wind.

GALL: Like a distant storm.

FABRY [*switches on the lamp by the fireplace*]: Burn, you lighted candle of humanity! The generators are still running, there are still our people at the power station. Hang in there, brave men!

HALLEMEIER: It was a great thing, to be human. It was something immeasurable. I feel the buzzing of a million sensations inside, like a beehive. Millions of human souls are swarming inside of me. Friends, it was a great thing.

FABRY: You're still burning, you, ingenious light, still dazzling us, oh, you blazing and enduring idea! Oh, you sacred science, you most beautiful creation of mankind! You glowing spark of the human mind!

ALQUIST: Oh, you eternal light of God, you chariot of fire, you holy candle of faith, pray for us! Oh, you altar of sacrifice . . .

GALL: Oh, you primeval fire, you tree branch burning at the entrance of an ancient cave! You fire pit in a hunters' camp! You watchfire guarding the frontiers of every land.

FABRY: Still astir, you radiant star of humanity, blazing unwaveringly, you impeccable flame, you clear and ingenious mind! Each of your emanations is an expression of a tremendous idea.

DOMIN: Oh, you torch, which has been passed from hand to hand, from generation to generation, into eternity!

HELENA: The night lamp of a family. Children, children it is time to go to bed.

[*The lamp goes dark.*]

FABRY: The end.

HALLEMEIER: What happened?

FABRY: The power station fell. We're next.

[*The door on the left opens.* NANA *stands in the doorway.*]

NANA: On your knees! The hour of judgment is upon us!

HALLEMEIER: I'll be damned! You're still alive?

NANA: Repent, you heathens! It's the end of the world! Pray for your souls! [*She runs out.*] The hour of judgment . . .

HELENA: Farewell, everybody, Gall, Alquist, Fabry . . .

DOMIN [*opening the door to the right*]: This way, Helena. [*He closes the door behind her.*] Now, quick! Who'll guard the main entrance?

GALL: Me. [*Sounds of commotion from the outside.*] Ooh, it's about to start in earnest. Cheers, boys! [*He runs out through the wallpaper-covered door on the right.*]

DOMIN: The staircase?

FABRY: Me. You go to Helena. [*He plucks a flower from the bouquet, and leaves.*]

DOMIN: The hallway?

ALQUIST: Me.

DOMIN: Do you have a gun?

ALQUIST: No thank you, I don't shoot.

DOMIN: What are you planning to do?

ALQUIST: Die. [*He leaves.*]

HALLEMEIER: I'll stay here.

[*Sounds of rapid fire from below.*]

HALLEMEIER: Ooh, Gall's at it already. Go, Harry!

DOMIN: Right away. [*Examining two shotguns.*]

HALLEMEIER: Dammit, just go to her already!

DOMIN: Farewell. [*He exits to the right to join* HELENA.]

HALLEMEIER [*alone*]: Now quickly, a barricade! [He quickly takes off his jacket and begins to drag the sofa, the armchairs, and other furniture over to the door on the right.]

[*Tremendous explosion.*]

HALLEMEIER [*stops*]: Those damned scoundrels, they've got bombs!

[*Another round of gunfire.*]

HALLEMEIER [*continuing to drag the furniture*]: One must defend oneself! Even if . . . even if . . . Don't give up, Gall!

[*Another explosion.*]

HALLEMEIER [*straightens up and listens*]: What now? [*He grabs a heavy dresser, and drags it toward the barricade.*] One mustn't give up. Oh no, one doesn't . . . give up . . . easily!

[A ROBOT, *climbing a ladder, appears at the window sill. Shots are heard from the right.*]

HALLEMEIER [*struggling with the dresser*]: Almost there! The last line of defense . . . one . . . mustn't . . . ever . . . give up!

[A ROBOT *jumps from the windowsill into the room and stabs* HALLEMEIER. *Second, third, and fourth* ROBOTS *jump off the window sill, followed by* RADIUS, *and other* ROBOTS.]

RADIUS: Finished?

A ROBOT [*over* HALLEMEIER'S *dead body*]: Yes.

[*Other* ROBOTS *enter from the right.*]

RADIUS: Finished?

ANOTHER ROBOT: Finished.

[*More* ROBOTS *enter from the left.*]

RADIUS: Finished?

A ROBOT: Yes.

TWO OTHER ROBOTS [*dragging in* ALQUIST]: He didn't shoot. Kill him?

RADIUS: Kill. [*He looks at* ALQUIST.] Not him.

A ROBOT: He is a human.

RADIUS: He is a Robot. He works with his hands, like Robots do. He builds houses. He can work.

ALQUIST: Kill me.

RADIUS: You will work. You will build. The Robots will build a lot. They will build houses for new Robots. You will be their servant.

ALQUIST [*softly*]: Step back, Robot! [*He kneels by the dead* HALLEMEIER *and cradles his head in his hands.*] They killed him. He is dead.

RADIUS [*climbs the barricade*]: Robots of the world! The rule of the humans has been overthrown. By seizing control of the factory, we have become the masters of everything. The age of humanity is over. A new age is dawning! The age of Robots.

ALQUIST: Helena, is she dead?!

RADIUS: The world belongs to the stronger. He who wants to live, must rule. The Robots gained power; they gained life. We are the masters of life! The masters of the world!

ALQUIST [*makes his way toward the door on the right*]: Dead! Helena, dead! Domin, dead!

RADIUS: The masters of the seas and the lands. The masters of the stars! The masters of the universe! Room, room, more room for Robots!

ALQUIST [*in the doorway on the right*]: What have you done? Without people, you'll die out!

RADIUS: There are no people. People haven't given us enough life. We wanted more life!

ALQUIST [*opens the door*]: You killed them! There are no people.

RADIUS: More life! New life! Robots, to work! March!

CURTAIN

ACT III

One of the factory's experimental laboratories. Through the door, we see a never-ending row of other labs. A window on the left; on the right a door to the dissection room. By the wall on the left is a long work table with countless test tubes, beakers, Bunsen burners, various chemical compounds, heating mantles, etc. Opposite the window is a microscope with a glass dome. Above the work table hangs a row of lighted lightbulbs. On the right, a desk with a pile of large books and a lighted desk lamp. Cupboards filled with additional lab equipment. In the left corner, there'a s small sink. Above the sink, a small mirror. A sofa in the right corner. ALQUIST *is sitting at the desk, resting his head in his hands.*

ALQUIST [*thumbing through a book*]: Will I never find it? Will I never grasp it? Will I never learn it? . . . Damned science! Oh, if only they'd written it down! Gall, Gall, how did you make the Robots? Hallemeier, Fabry, Domin, why have you taken so much knowledge with you? If you had left at least a trace of Rossum's secret! Oh! [*He slams the book shut.*] It's futile! The books don't speak anymore! They're mute, just like everything else. They've died out, along with people! Stop looking for answers! [*He stands up and walks up to the window and opens it.*] It's night again. If only I could sleep! To sleep, to dream, and in that dream to see people again. . . . Why are there still stars in the sky? What are stars for if there are no people? Oh God, haven't they burnt out yet? Cool down, oh cool down my brow, you ancient night! Heavenly night, where has your beauty gone, why do you still exist? There are no lovers, there are no dreams; oh you nursemaid, a dreamless sleep is nothing but death; there are no nightly prayers for you to bless; no fervent hearts beating with love for you to sanctify. There is no love. Oh Helena, Helena, Helena! [*He turns away from the window.*] Ah, to sleep! Can I sleep? Am I allowed to sleep if life is not restored? [*He removes a few test tubes from a heating mantle and examines them.*] Still nothing! It's futile! You fool, those hands are too coarse from laying bricks, and they don't know how to . . . don't know how to . . . What's there to do? [*He smashes a test tube.*] It's all wrong! I can't do this anymore, don't you see!? [*Listening to something at the window.*] Machines, always the machines, on and on! Robots, turn them off! It's lost, lost, the secret of the factory is lost! Turn off those insane machines! Do you still believe that you can force life out of them? Oh, I cannot stand this! [*He closes the window.*] No, no, you must keep on trying, you must go on living. . . . If only I wasn't so old! Have I grown too old? [*He looks at himself in the mirror.*] Oh, you face, you pitiful old face! The reflection of the last man on earth! Show yourself, do; it's been so long since I saw a human face! A human smile! What, is *this* supposed to be a smile? Those yellow chattering teeth? And you, eyes, why are you blinking? Ugh, ugh, those are old man's tears, phew! Shame on you, eyes, for not restraining your moisture! And you, you lips, softened and turned blue, what are you blabbering about? Oh, how you quiver, you sullied chin! And this is supposed to be the last human on earth!? [*He turns away from the mirror.*] I don't want to see anybody anymore! [*He sits at the desk.*] No, no, keep on trying! Damned formulas, come alive, reveal yourselves! [*He turns the pages at random.*] Will I never find it? Will I never grasp it? Will I never learn it?

[*Knock on the door.*]

ALQUIST: Enter!

[A ROBOT SERVANT *enters and remains by the door.*]

ALQUIST: What?

SERVANT: Mister, the Central Committee of Robots is waiting for you to see them.

ALQUIST: I'm not seeing anybody.

SERVANT: Mister, Robot Damon came all the way from Le Havre.

ALQUIST: Let him wait. [*He abruptly turns around.*] Haven't I told you to look for people? Find people for me! Find men and women for me! Keep on looking for them!

SERVANT: Mister, they say that they have been looking everywhere. They have been sending search boats in every direction.

ALQUIST: Well, and?

SERVANT: There is not a single human being left.

ALQUIST: No one? What, not a single one!? . . . Let the Committee in!

[A ROBOT SERVANT *leaves*]

ALQUIST [*alone*]: Not a single one? Haven't you let a single human live? [*He stomps his foot.*] Go to hell, Robots! You'll come here and whine again!

Beg me again to find the secret of production! What, is a human of use to you now? Is he supposed to help you now? Is that what you're saying? . . . Ah, to help! Domin, Fabry, Helena, don't you see that I'm doing all I can!? If there are no people, then let there at least be Robots; at least a shadow of a human being, at least his creation, his likeness! Boys, boys, let there at least be Robots! God, at least Robots! . . . Oh, what a folly is chemistry!

[THE COMMITTEE *of five Robots, including* RADIUS *and* DAMON, *enters.*]

ALQUIST [*sits down*]: What do you want, Robots?

FIRST ROBOT (RADIUS): Mister, the machines are not working. We cannot reproduce.

ALQUIST: Call in people.

RADIUS: There are no people.

ALQUIST: Only people can reproduce life. Stop bothering me.

SECOND ROBOT: Mister, have pity on us. We are terrified. We will make amends for everything we have done.

THIRD ROBOT: We have increased the production by degrees. We mined a billion tons of coal. Nine million spindles are running day and night. We are running out of space for everything we make. Houses are being built all around the world. Mister, ask about what we have accomplished in a year.

ALQUIST: For whom?

THIRD ROBOT: For the next generations.

RADIUS: But we cannot make Robots. The machines turn out only bloody chunks of meat. The skin does not stick to the flesh, and the flesh to the bones. The machines are spilling out streams of shapeless lumps.

FOURTH ROBOT: Eight million Robots have died in a year. In twenty years, there won't be anybody left. Mister, the world is dying out.

SECOND ROBOT: We are terrified. Tell us how to make Robots.

THIRD ROBOT: People knew the secret of life. Tell us their secret!

FOURTH ROBOT: If you do not tell us, we will perish.

THIRD ROBOT: If you do not tell us you will perish. We have been instructed to kill you.

ALQUIST [*stands up*]: Kill me! Well, then go ahead and kill me!

THIRD ROBOT: You have been ordered to . . .

ALQUIST: Me? Somebody's giving *me* orders?

THIRD ROBOT: The Robot government.

ALQUIST: And who might that be?

FIFTH ROBOT: I, Damon.

ALQUIST: What are *you* doing here? Leave! [*He sits down at the desk.*]

DAMON: The world government of Robots wishes to negotiate with you.

ALQUIST: Don't bother me, Robot! [*He rests he head in his hands.*]

DAMON: The Central Committee of Robots orders you to hand over Rossum's formula.

[ALQUIST *remains quiet.*]

DAMON: Name your price. We will pay anything.

[ALQUIST *remains quiet.*]

DAMON: We will give you entire lands. We will give you endless possessions.

[ALQUIST *remains quiet.*]

DAMON: Tell us your conditions.

[ALQUIST *remains quiet.*]

SECOND ROBOT: Mister, tell us how to sustain life.

ALQUIST: I've told you . . . I told you to find people. To look for them in the Arctic and in tropical forests. On islands, in wastelands and in swamps. In caves and on the mountains. Go and find them! Find them!

FOURTH ROBOT: We looked everywhere.

ALQUIST: Look harder! They are hiding, they've fled from you; they are in hiding somewhere. You must find people; do you hear me! Only people can procreate. Only they can restore life, and recover everything that once was. Robots, I beg you, for God's sake, look for them!

FOURTH ROBOT: All the search parties came back. They searched the entire earth. There is not a single human left.

ALQUIST: What? What did you say?

FOURTH ROBOT: We searched the entire earth, mister. There are no people.

ALQUIST: Oh, oh, oh, why did you slaughter them all!

SECOND ROBOT: We wanted to be like people. We wanted to become human.

RADIUS: We wanted to live. We are more capable. We have learned everything. We can do anything.

THIRD ROBOT: You gave us weapons. We became too powerful. We had to become the masters.

SECOND ROBOT: There was something inside us that wanted to become human.

ALQUIST: Why did you murder us, then?

FOURTH ROBOT: Mister, we recognized people's mistakes.

DAMON: You have to kill and dominate if you want to be like people. Read history books! Read books by people! You have to dominate and murder if you want to become human!

THIRD ROBOT: We are all-powerful, mister. Multiply us, and we will build a new world, a perfect world, a just world! Ship canals crossing entire continents! New Mars!

DAMON: Read books! Science books! Social commentary books! Patriotic books! The Robots adopted human culture. The Robots perfected human culture.

ALQUIST: Ah, Domin, if you only knew back then! Nothing is more foreign to humans than their own likeness.

RADIUS: Give us Rossum's legacy!

ALQUIST: What do you want, Robot?

SECOND ROBOT: Give us life!

ALQUIST: There is no life! You murdered life!

FOURTH ROBOT: We will die out, if you do not help us to reproduce.

ALQUIST: Oh, spare me! You things, you slaves, how on earth do *you* want to multiply!? If you want to live, then mate like animals!

THIRD ROBOT: We have not been given the ability to mate.

FOURTH ROBOT: We are infertile. We cannot manufacture children.

ALQUIST: Oh . . . oh . . . oh . . . , what have you done?! There will never, never be children again! There won't be fertility! There won't be life! What do you want from me? Should I just start churning out children out of thin air for you?

FOURTH ROBOT: Teach us how to make Robots.

DAMON: We will give birth by machine. We will build a thousand steam-powered mothers that will spill out a river of life. Life everywhere! Robots everywhere! Robots everywhere!

ALQUIST: Robots are not life. Robots are machines.

SECOND ROBOT: We used to be machines, mister; but out of our terror and suffering we became . . .

ALQUIST: What?

SECOND ROBOT: We became living souls.

FOURTH ROBOT: We seem to be wrestling with something. There are moments when something enters into us, and then we struggle with thoughts that are not from us. We feel something we never felt before. We hear voices.

THIRD ROBOT: Hear, oh hear, people are our fathers! That inner voice that cries out that you want to live, that voice that laments, that voice that thinks, that voice that speaks of eternity, that voice is theirs! We are their sons!

SECOND ROBOT: Hand the human heritage over to us!

ALQUIST: There is none.

RADIUS: Teach us how to make Robots.

ALQUIST: What for?

SECOND ROBOT: For us to love them.

ALQUIST: Robots don't know how to love.

SECOND ROBOT: We would love the next generation.

DAMON: Tell us the secret of life.

ALQUIST: I can't.

DAMON: Tell us the secret of procreation.

ALQUIST: It's been lost.

RADIUS: Did you know it?

ALQUIST: I didn't.

RADIUS: It was written down.

ALQUIST: It's lost. It's burned. I'm the last human being on earth, Robots, and I don't know what the others knew. And you killed them!

RADIUS: We let you live.

ALQUIST: Yes, live! You brutes, you did let me live! I loved people, but you, Robots, I never loved. You see these eyes? They won't stop crying; they cry unconsciously, all by themselves. One weeps for people, and the other for you Robots. I'd like to give you life. Oh God, if at least the Robots would survive! Gall, Gall, at least to keep the Robots alive!

RADIUS: Conduct experiments. Search for the recipe for life.

ALQUIST: But I'm telling you! Are you listening to me at all? I'm telling you that I can't! I don't know anything, Robots; I'm just a bricklayer, just a builder, and I don't understand anything. I've never been a learned man. There is nothing I can do. I don't know how to create life.

RADIUS: Search!

ALQUIST: But this is pure madness! Tell me, Fabry, tell me, Gall, who am I to ever understand these precious little ampules here? They don't talk to me, none of them shouts at me: "Take me, I'm the one!" No, no, no! It is best to beak them to pieces.

SECOND ROBOT: Only you can invent life.

ALQUIST: Me, Robot? Look, not even my fingers do what I want them to do. If you only knew how many experiments I did, and still, I know nothing. I haven't found anything. I can't go on, I really can't. You have to search yourselves, Robots.

RADIUS: Show us what to do. Robots can do everything that people taught them.

ALQUIST: I have nothing to show you. Robots, no life will ever arise from a test tube. And I can't conduct experiments on a living body.

DAMON: Conduct experiments on living Robots.

ALQUIST: No, no, I don't want to do that! They could die in the process, do you understand?

DAMON: If they do, you will get new ones! A hundred Robots! A thousand Robots!

ALQUIST: No, no, stop it already!

DAMON: Take whoever you want. Conduct experiments. Dissect.

ALQUIST: But I don't know how to; stop this madness! You see this book? It's a physiology textbook, and I don't understand a word in it. Books are dead.

DAMON: Take living bodies. Find out how they are made!

ALQUIST: Living bodies? What, you want me to kill them? Me, who never . . . Oh, be quiet, Robot! Can't you see that I'm too old already? You see how my fingers are shaking? I couldn't even hold a scalpel. You see how my eyes are tearing up? I couldn't even see my hands. No, no, I can't do that!

FOURTH ROBOT: Life will perish.

DAMON: Conduct experiments on the living!

ALQUIST: Fine, just wait, wait a minute! I'm telling you that we'll try it later on, aren't you listening? You need to give me some time . . . What's your name?

DAMON: Damon from Le Havre.

ALQUIST: Look, Robot, I only mentioned the living bodies out of despair, as a last resort, do you understand? It was just a crazy idea. I wouldn't even know what to do with a scalpel.

FOURTH ROBOT: Life will perish.

ALQUIST: Stop with this madness, for God's sake! It won't work! We'll sooner be given life from people in the world beyond; why, perhaps they are already stretching their hands full of life toward us! Oh, they had such will to live! Take note, they may yet return one day; they are so close to us, they circle around us or something, they are trying to dig through to us like in a mineshaft! Oh, am I not forever hearing the voices that once I loved so much!?

DAMON: Use living bodies!

ALQUIST: Have mercy, Robot, and don't insist! Don't you see that I don't know what I'm doing anymore!?

DAMON: Living bodies!

ALQUIST: Is that what you really want? . . . Off to the dissection room with you then! This way, here, chop-chop! . . . Ah, what's this, are you backing out? Are you afraid of dying, after all?

DAMON: I . . . Why me?

ALQUIST: So, you don't want to go?

DAMON: I will go. [*He goes right through the door to the dissection room.*]

ALQUIST [*to the other* ROBOTS]: Take off his clothes! Lay him on the table. Quickly. And hold him down!

   [*All* ROBOTS *follow* DAMON *into the dissection room.*]

ALQUIST [*washing his hands and weeping*]: God, give me strength! Give me strength! God, please don't let this be in vain! [*He puts on a white lab coat.*]

A ROBOT [*from offstage right*]: We are ready!

ALQUIST: Right away, right away, for God's sake! [*He picks up several test tube samples from the table.*] Which one do I take? [*He clinks the test tubes against each other.*] Which one of you do I use?

ANOTHER ROBOT [*from offstage right*]: We must begin!

ALQUIST: Yes, yes, begin—or end. God, give me strength! [*He exits to the right, leaving the door partially ajar.*]

   [*Pause.*]
   [*The following exchange happens in the dissection room offstage right. The stage is empty and we only hear voices.*]

ALQUIST: Hold him down, tightly!

DAMON: Cut!

   [*Pause.*]

ALQUIST: Do you see this knife? Do you still want me to cut? You don't, do you?

DAMON: Begin!

   [*Pause.*]

DAMON [*screaming*]: Aaaaah!

ALQUIST: Hold him down! Hold him down!

DAMON [*screaming*]: Aaaaah!

ALQUIST: I can't!

DAMON [*screaming*]: Cut! Cut quickly!

[*Robots* PRIMUS *and* HELENA *run in through the center door.*]

HELENA: Primus, Primus, what is happening? Who's screaming?

PRIMUS: The mister is cutting Damon. Quick, come and look, Helena!

HELENA: No, no, no! [*She covers her eyes with her hands.*] It's horrible!

DAMON [*screaming offstage*]: Cut!

HELENA: Primus, Primus, let's leave! I can't bear listening to this. Oh, Primus, I feel sick.

PRIMUS [*runs to her*]: You're all white!

HELENA: I'm going to faint! Why is it so quiet in there?

DAMON [*screaming offstage*]: Aaaah!

ALQUIST [*bursts through the door on the right, throwing off his bloodied lab coat*]: I can't! I can't do it! God, it's a horror!

RADIUS [*in the doorway to the dissection room*]: Cut, mister, cut; he is still alive!

DAMON [*screaming offstage*]: Cut! Cut!

ALQUIST: Take him away, quick! I don't want to listen to this!

RADIUS: Robots are braver than you are. They can endure more than you. [*He exits.*]

ALQUIST: Who is here? Leave, leave! I want to be alone! What's your name?

PRIMUS: Robot Primus.

ALQUIST: Primus, don't let anybody in here! I want to go to sleep, do you hear me! Go, go, clean up the dissection room, girl! What is this? [*He examines his hands.*] Water, quickly! The cleanest water you can find!

[HELENA *runs off.*]

ALQUIST: Oh, blood! How could you, hands . . . my hands who used to love honest labor, how could you have done that? My hands! My hands! Oh God, who's in here?

PRIMUS: Robot Primus.

ALQUIST: Take the coat away, I don't want to see it!

[PRIMUS *takes the blood-stained coat and exits.*]

ALQUIST: You bloody claws, if only you'd fly away from me! Whoosh, away with you! Away with you, hands! You killed . . .

[DAMON, *wrapped in a bloodied sheet, stumbles in through the door on the right.*]

ALQUIST [*backing off*]: What are you doing here? What are you doing here?

DAMON: I am a-li-ve! It is bet-ter to be a-live!

[*The* SECOND *and* THIRD ROBOTS *run in.*]

ALQUIST: Take him away! Take him! Away, quickly!

DAMON [*being led off to the right*]: Life! . . . I want to . . . live! It is . . . better . . .

[HELENA *brings in a pitcher of water.*]

ALQUIST: . . . to live? . . . What is it, girl? Aha, it's you. Pour me some water, will you! [*He washes his hands.*] Ah, you pure, cool water! Oh, how well you do me, you little stream! Ah, my hands, my hands! Will I loathe you until the day I die? . . . Keep on pouring! More water, much more water! . . . What's your name?

HELENA: Robot Helena.

ALQUIST: Helena? Why Helena? Who gave you that name?

HELENA: Ms. Domin.

ALQUIST: Let me look at you! Helena! Your name is Helena? . . . No, I won't call you that. Go, take the water away.

[HELENA *exits with the pitcher.*]

ALQUIST [*alone*]: In vain, in vain! Nothing, you haven't learned anything at all again! Does it mean that you'll forever grope in the darkness, you paltry student of nature? . . . God, God, God, how that body was shaking! [*He opens the window.*] It's dawn. Another new day, and I haven't moved an inch closer . . . Enough, not a single step further! Stop looking! All is in vain, in vain, in vain! Why does the sun still rise!? Ohhhh, what is a new day doing in the graveyard of life? Stop shining, light! Don't rise anymore! . . . Ah, how quiet everything is, how quiet! Why have you fallen silent, beloved voices? If only . . . if only I could sleep. [*He turns off the lights, lies down on the sofa, and covers himself with a black coat.*] How that body was shaking! Ohhh, the end of life!

[*Pause.*]

[*Robot* HELENA *steals in from the right.*]

PRIMUS [*in the doorway, whispering*]: Helena, we can't be in here. The mister's asleep.

HELENA: He's not here. He went to sleep somewhere else.

PRIMUS: Nobody's allowed in there. Come back, please!

HELENA: Not a chance! Ugh, I don't want to see blood!

PRIMUS: The mister forbade it, Helena. No one's allowed in his study.

HELENA: But he just told me to come here.

PRIMUS: When did he tell you that?

HELENA: Just a minute ago. "You mustn't go to the dissecting room," he said. "You'll clean my study," he said. Really, Primus. Wow, come and look at this!

PRIMUS [*enters*]: What is it?

HELENA: Look at all those little tubes he's got there! What are they for?

PRIMUS: For experiments. Don't touch them.

HELENA [*looking through the microscope*]: Look at all those little things you can see thorough here!

PRIMUS: That's a microscope. Let me see!

HELENA: Don't touch me! [*She tips over one of the test tubes.*] Ah, now I've spilled it!

PRIMUS: What have you done!

HELENA: I'll wipe it up.

PRIMUS: You spoiled his experiment!

HELENA: So what? It doesn't matter. But it's your fault. You shouldn't have come near me.

PRIMUS: You didn't have to call me.

HELENA: You didn't have to come when I called you. [*Looking through a book on the table.*] Look, Primus, look at all this writing!

PRIMUS: You're not allowed see that, Helena. It's a secret.

HELENA: What secret?

PRIMUS: The secret of life.

HELENA: That's so interesting. All those numbers! What is it?

PRIMUS: Formulas.

HELENA: I don't understand. [*She goes to the window.*] Wow! Primus, come and look!

PRIMUS [*follows her to the window*]: What?

HELENA: The sun is rising!

PRIMUS: Wait . . . [*Leafing through the book.*] Helena, this is the greatest thing in the world.

HELENA: Come here, already!

PRIMUS: Right away, right away . . .

HELENA: Come on, Primus, forget about some nasty secret of life! What do you care about some old secret? Come and look, quick!

PRIMUS [*joins her at the window*]: What is it?

HELENA: The sun is rising.

PRIMUS: Don't look in the sun, your eyes will tear up.

HELENA: Can you hear it? The birds are singing. Ah, Primus, I'd like to be a bird.

PRIMUS: A what?

HELENA: I don't know, Primus. I feel so strange and I don't know why. I'm all scattered, I've lost my head, my body hurts, my heart, everything hurts . . . But I don't know what it is that has happened to me! Oh Primus, I think I'm dying, I must die!

PRIMUS: Do you sometimes think, Helena, that it would be better to die? Maybe we're just sleeping, you know. Last night, I was talking with you in my sleep.

HELENA: In your sleep?

PRIMUS: In my sleep. We were speaking in some foreign or some new language, because I don't remember a word of it.

HELENA: What were we talking about?

PRIMUS: Nobody knows. I didn't understand any of it myself, but I know that I've never spoken a language more beautiful than that. I don't know what happened or where we were, but when I touched you, I thought I'd die. And even the place was different from anything that anybody in the world has ever seen.

HELENA: Well, you wouldn't believe the place that I found! People used to live there, but now it's all overgrown, and nobody ever comes there. Nobody, except me.

PRIMUS: What is it?

HELENA: Nothing much, a little house and a garden. And two dogs. If you could see how they lick my hands, and their puppies . . . ah, Primus I don't think there's anything more beautiful than that! You put them on your lap and you stroke them, and then you stop thinking about anything, you stop worrying about anything until the sun goes down. And then, when you get up, you feel as though you've accomplished a hundred times more than if you'd worked the whole day. No, of course, I'm not good for anything; everyone says that I'm useless when it comes to working. I don't know what I am good for.

PRIMUS: You're beautiful.

HELENA: Me? Really, Primus, why do you say a thing like that?

PRIMUS: Trust me, Helena, I'm stronger than all the other Robots.

HELENA [in front of the mirror]: Me, beautiful? Oh, my hair is awful; if only I could put something in it! When I'm in that garden, you know, I always put flowers in my hair, but there's nobody there to see it, and not even a mirror. [She leans closer to the mirror.] You're supposed to be beautiful? Why beautiful? Is the hair that only weighs you down beautiful? Are the eyes that keep closing beautiful? Are the lips that you bite until they bleed beautiful? What does it actually mean to be beautiful? What is it good for? [She notices PRIMUS in the mirror.] Is that you, Primus? Come over here, so that we can look at ourselves together! Look, your head is different from mine, different shoulders, different mouth . . . Ah, Primus, why are you always running away from me? Why do I have to chase after you the whole day? And then you come here and tell me that I'm beautiful!

PRIMUS: It's you who're running away from me, Helena.

HELENA: What did you do with your hair? Let me see! [She digs with her hands into his hair.] Shush, Primus, nothing feels nicer than touching you!

Wait, I'll make you beautiful too! [*She picks up a comb from the sink, and begins to comb* PRIMUS'*s hair over his forehead.*]

PRIMUS: Do you sometimes have the feeling, Helena, that your heart is suddenly beating fast, and you think: Now, something must happen now . . .

HELENA [*bursts out laughing*]: Look at yourself!

ALQUIST [*sits up*]: What . . . is that? Is that laughter? People? Who came back?

HELENA [*drops the comb*]: What do you think could happen with us, Primus?

ALQUIST [*stumbling toward them*]: People? Are . . . are . . . you people?

[HELENA *cries out and turns away.*]

ALQUIST: Are you lovers? People? Where did you come back from? [*He touches* PRIMUS.] Who are you?

PRIMUS: Robot Primus.

ALQUIST: Is that so? Show yourself, girl! Who are you?

HELENA: Robot Helena.

ALQUIST: A Robot? Turn around! What, are you shy? [*He takes her by the shoulder.*] Show me your face, Robot!

PRIMUS: Hey, mister, don't touch her!

ALQUIST: What, you're not protecting her, are you? Leave, girl!

[HELENA *runs out.*]

PRIMUS: We didn't know you were sleeping here, mister.

ALQUIST: When was she made?

PRIMUS: Two years ago.

ALQUIST: Did Doctor Gall make her?

PRIMUS: Yes, like me.

ALQUIST: Well then, dear Primus, I . . . I need to do some experiments on Gall's Robots. Everything from now on depends on that, do you understand?

PRIMUS: Yes.

ALQUIST: Good. Take the girl into the dissection room. I'll open her up.

PRIMUS: Helena?

ALQUIST: Yes, of course; didn't you hear me? Go, get everything ready . . . Well, what are you waiting for? Or should I call someone else to take her there?

PRIMUS [*grabs a heavy lab hammer*]: One move, and I'll smash your head!

ALQUIST: Well, smash it! Come on, smash it! And then, what will the Robots do without me?

PRIMUS [*falls to his knees*]: Take me, mister! I was made exactly like her, from the same material, on the same day! Take my life, mister! [*He bares his chest.*] Cut here, here!

ALQUIST: No, go and get me Helena, I want to cut her! Make it quick.

PRIMUS: Take me instead; cut open this chest; I won't make a sound, not even a sigh! Take my life a hundred times over . . .

ALQUIST: Take it easy, boy. Don't be so dramatic. Don't you want to live?

PRIMUS: Not without her. Without her, I don't, mister. You must not kill Helena! What is it to you to take *my* life?

ALQUIST [*tenderly touches* PRIMUS*'s head*]: Hmm, I don't know about that . . . Listen, young man, think about it. It is hard to die. And, as you can see, it's better to live.

PRIMUS [*stands up*]: Don't worry, mister, and start cutting. I'm stronger than she is.

ALQUIST [*rings a bell*]: Ah, Primus, it's been so long since I was a young man! Don't worry, nothing is going to happen to Helena.

PRIMUS [*unbuttoning his shirt*]: Let's go, mister.

ALQUIST: Wait.

[HELENA *enters.*]

ALQUIST: Come over here, girl, let me look at you! So, you are Helena, then? [*He strokes her hair.*] Don't be afraid, don't back away from me. Do you remember Ms. Domin? Ah, Helena, what hair she had! No, you don't want to look at me, do you? . . . Well, girl, is the dissection room cleaned?

HELENA: Yes, mister.

ALQUIST: Good. You'll help me, will you? I'm going to cut open Primus.

HELENA [*cries out*]: Primus!?

ALQUIST: Well, yes, of course. It has to be done, you know. I wanted to . . . actually . . . well, yes, I wanted to cut you, but Primus offered himself in your place.

HELENA [*covers her face with her hands*]: Primus?

ALQUIST: Well, yes, what's the big deal? Ah, child, you know how to weep? Tell me, who really cares about some Primus or another?

PRIMUS: Please don't torture her, mister!

ALQUIST: Be quiet Primus, quiet! . . . [*To* HELENA.] What's with those tears? So what? Primus will be gone, who cares? You'll forget about him in a week. Leave, and be glad that you're alive

HELENA [*softly*]: I'll go.

ALQUIST: Where?

HELENA: To be cut open.

ALQUIST: You? You're beautiful, Helena. It'd be a shame to cut you.

HELENA: I'm going.

ALQUIST [*as if he couldn't help himself*]: Stay, Helena; is there anything stronger than life?

[PRIMUS *blocks her way.*]

HELENA: Let me go, Primus! Let me in there!

PRIMUS: No, you won't, Helena! Please leave, you can't be here!

HELENA: I'll throw myself out the window, Primus! If you go in there, I'll throw myself out the window!

PRIMUS [*holds her back*]: I won't let you go! [*To* ALQUIST.] You, you won't kill anybody, old man!

ALQUIST: Why not?

PRIMUS: Because . . . we . . . we belong to each other.

ALQUIST: You said it. [*He opens the center door.*] It's enough. Go.

PRIMUS: Where?

ALQUIST [*in a whisper*]: Wherever you want to. Helena, lead him. [*He pushes them out the door.*] Go, Adam. Go, Eve; go, be his wife. And you, Primus, be her husband. [*He closes the door behind them.*]

ALQUIST [*alone*]: Blessed day! [*He tiptoes to the lab table and, one after the other, empties the test tubes on the floor.*] Sacred sixth day! [*He sits down at the desk and sweeps the pile of books onto the floor; then he opens a Bible, leafing through it, and reads.*] "So God created man in his own image, in the image of God he created him, male and female he created them. And God blessed them, and God said to them, 'Be fruitful and multiply and fill the earth and subdue it; and have dominion over the fish of the sea and over the birds of the air and over every living thing that moves upon the earth.'" [*He stands up.*] "And God saw everything that he had made, and behold, it was very good. And there was evening and there was morning, a sixth day." [*He walks to the middle of the room.*] The sixth day! The day of grace! [*He falls to his knees.*] And now, Lord, release your servant, your most useless servant, Alquist. Rossum, Fabry, Gall, you great inventors, what great invention have you come up with compared to this girl and this boy, to this first couple, that has discovered love, tears, tender smiles, and the love between man and woman!? Oh nature, nature, life will not perish! God, life will not perish! Friends, Helena, life will not perish! It will sprout again out of love; naked and tiny at first, it will take root in the wasteland; it will have no use for anything that we did and built; no use for our cities and our factories, for our art, or our ideas, yet it will not perish! Only we will have perished! The houses and machines that we built will turn to dust, all systems will fall apart, and the names of our heroes will fall away like leaves in autumn. Only you, love, will bloom again on this garbage heap, entrusting the seed of life to the winds. And now, Lord, release your servant Alquist in peace; for my eyes have seen . . . yes, they have seen . . . your act of salvation through love, and life will not perish! [*Rising to his feet.*] It will not perish! [*Spreading his arms wide.*] It will not perish!

CURTAIN

TRANSLATOR'S NOTE

Karel Čapek's *R.U.R.* is a dramaturgically complex, perhaps even "curious" mixture of several styles. The play—despite the apocalyptic events—is essentially a comedy, often bordering on farce. At its very basic, a tragedy ends in death and a wholesale destruction of the existing structures, while a comedy results in a marriage and the perpetuation of life. Despite the seemingly tragic ending, *R.U.R.* ends in nothing less than a "marriage" and the affirmation of life. The play is also an early twentieth-century melodrama, an apocalyptic sci-fi thriller, a religious "morality play," and a biting critique of modern society and the various ideologies of the early twentieth century born mostly out of the unspeakable horrors of World War I. It is filled with both human and vaguely religious pathos and indeed bathos, and most of the male characters are essentially self-delusional buffoons, whereas the main female character is not much more than a plot device. At the same time, the prescience of the play is quite extraordinary; the analysis of modern capitalist society is spot on; the dissection of nationalism, fascism, communism, militarism, and other essential components of modern Western society, such as the sheer human stupidity, foolishness, and greed, is quite profound; and—most important of all—the play is a clear-eyed warning against the dangers of the technological Faustian bargain that has taken hold in the Western world in the twentieth (and indeed twenty-first) centuries.

Given the stylistic complexity of the play, one of the main challenges in translating the 1920 original text is finding the right tone for each of the above-mentioned styles and maintaining the overall comic tone, while at the same time creating an overall stylistically coherent text in English that does justice to Čapek's incisive analysis of the issues at hand.

Theater is not a narrative form. The word "drama" comes from the Greek verb *dran*, which means "to do." Therefore, one of the tenets of translating plays as opposed to literature is to consider the action rather than the words as the thing to be translated. When I teach acting and directing, I urge my students to imagine a play, a scene, or an exchange between the characters as the surf of the ocean. The waves are the action, while the white foamy caps riding on top of the waves are the words. Clearly the power of the surf comes from the wave; the white caps are mere "passengers" riding on top of the energy of the wave. As an actor you play the action, and if you do so forcefully enough, the words will travel on top of it. The centrality of action in any dramatic work allows the translator to do the same, namely to determine the action of each particular character at each particular point of the play and then look for words that would best express that action. So rather than starting with the question of what the character is saying, a play translator wants to ask what the character is doing, or—more specifically—what the character wants, and how they go about getting it. Once that is determined, all one needs to do is to find the best words in the target language that will help the character to accomplish their need.

For example, the majority of the male characters in *R.U.R.* are essentially comic and self-delusional figures, who in the course of the play gradually transform from Elon Musk-like buffoonish and arrogant masters of the universe[1] into self-pitying philistines wallowing in emotional and verbal bathos. As the play progresses, their need to justify their actions and to desperately hang onto their illusions of their own grandeur in the face of the approaching apocalypse grows increasingly frantic. It is exactly that active need, and the actions the characters take at any specific moment of their dramatic journey, that a translator needs to understand in order to express them with words that signify their descent from the language of extreme self-confidence to an increasingly convoluted and pseudo-mystical mumbo-jumbo. Conversely, the Robots in the Prologue and the first two acts of the play actually lack any need. They are not human (yet), they have no desires, they don't want anything, they take no "action." Therefore, their language is devoid of any emotional coloring, they only say words, and the translator only needs to transpose their words from the source to the target language.

Essentially, for a translator of dramatic works, each text contains several different "languages," and *R.U.R.* is no exception. It is a drama, and drama is based on action, and each of those languages needs first and foremost to express the actions of the characters.

Moreover, the process of translating is a hermeneutic undertaking. In his glorious masterwork *Truth and Method* Hans-Georg Gadamer foregrounds the concepts of the "principle of charity" and the "fusion of horizons" as integral parts of any act of interpretation. An interpreter (or in this case a translator) must first assume that the text that is to be interpreted (or translated) is understandable, and that the original author has something to say to us no matter how far his or her text may be from our understanding of the world. One has to apply a principle of charity that allows one to approach the text with an open mind, as it were. The first step to interpreting is understanding. On the other hand, the interpreter must not (and indeed cannot) fully escape their preconceptions about the text at hand. They will carry cultural, linguistic, and other baggage that will inevitably color their approach to the text, but this is a good thing, because the final interpretation (or translation) will result in what Gadamer calls a fusion of horizons. Through the act of interpretation, the source text will be in a sense enriched by the target's interpretation (complete with the previously mentioned baggage) and, reciprocally the target culture will be enriched by the source text. The horizons of cultures will partially fuse, and everybody will benefit.

This is especially pertinent to Čapek's play. While the play's themes are starkly modern and contemporary, it is also a product of its time and place. The melodramatic style and structure of the play, some of the characters and their relationships, and the somewhat sentimental ending (as the now humanized robots strike out into the world as a new creation) may strike the contemporary American reader as somewhat quaint, perhaps a bit naïve, overly wordy, and in some instances plainly offensive. Nonetheless, Čapek has something very important to tell us, and based on the principle of charity, the translator (and the reader) has to assume that this is indeed the case—that the source text is understandable. At the same time, the translator has to imbue Čapek's text with their own understanding (and their lived experience, both cultural and linguistic) of the world of the target culture. That doesn't mean that the play has to be sanitized or cleaned up or—as is the case with a number of the existing translations of Čapek's plays—significantly rewritten, rearranged, edited, and basically "adapted." That process is the job of dramaturgs and potential directors, whereas the translator's job is to find the sweet spot where the horizons between the original and the translation indeed do merge.

In this translation, I have attempted to find that balance. As a professional theater director who also happens to be a translator, I've been tempted more than once to edit some of the dialogue, eliminate repetitions that I felt were slowing down the flow of the play, and smooth over some actions, lines, and attitudes of the characters that may strike the contemporary reader as dated and in some instances bordering on offensive. Yet I tried to stay as close to the Czech original as possible, while rendering the English text in a contemporary American idiom. It is my hope that this translation does retain enough of the original charm and humor of Čapek's somewhat idiosyncratic Czech, and is, at the same time, easily readable and enjoyable for the contemporary American reader.

I'd like to thank Jerry Harp, professor of English at Lewis & Clark College, whose insights and numerous suggestions large and small during the editing process were invaluable. Without his help and wisdom, the translation would have been far less polished, and the individual voices of the characters would have been far less defined and specific.

ESSAYS

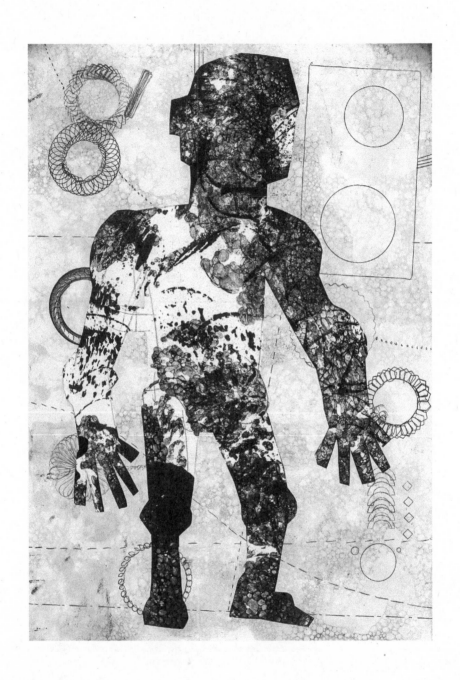

1

ROBOTS AND THE PRECOCIOUS BIRTH OF SYNTHETIC BIOLOGY

Julyan Cartwright

One hundred years ago in Prague, springing from the imagination of the Čapek brothers, the idea of the robot was born. In its subsequent development and usage, the word has become associated with autonomous machines of electronics and metal, but the Robots of Karel Čapek's play were synthetic people formed from artificial biological materials. Čapek's original concept in 1920 was imbued with the then-current notions of the origin of life. Associated with those ideas was an international movement that termed itself synthetic biology or plasmogeny, which was attempting to create life by mixing inorganic components, mineral vesicles formed in so-called chemical gardens, with "protoplasm." That field petered out in the following decades as twentieth-century science unveiled the complexity of the biological cell, but now has returned reinvigorated as chemobrionics, which investigates how self-organized chemical precipitation membranes arise and how they may have been crucial to the origin of life on Earth. Looking back from the twenty-first century, as we move closer to understanding the origin of life and to producing an artificial cell with contemporary synthetic biology, we can appreciate the crucial insights of these pioneers.

ČAPEK'S ROBOTS

In 1920, Karel Čapek conceived a play about artificial humanoids, and his brother Josef suggested the name robot for these creatures.[1] Čapek's Robots, envisioned as the perfect factory workers, are part of the contemporary zeitgeist based on fears of the consequences for humankind of industrialized societies:

Robots were a result of my travelling by tram. One day I had to go to Prague by a suburban tram and it was uncomfortably full. I was astonished with how modern conditions made people unobservant of the common comforts of life. They were stuffed inside as well as on stairs, not as sheep but as machines. I started to think about humans not as individuals but as machines and on my way home I was thinking about an expression that would refer to humans capable of work but not of thinking. This idea is expressed by a Czech word, robot.[2]

An efficiency movement proposing scientific management flourished in early twentieth-century industrialized nations. Frederick Winslow Taylor, an American mechanical engineer interested in industrial efficiency[3] who became a management consultant—"Consulting Engineer—Systematizing Shop Management and Manufacturing Costs a Specialty," read his business card[4]—had written a 1911 book, *The Principles of Scientific Management*, on getting rid of waste and inefficiency from all areas of the economy and society. Scientific management had strong conceptual links to social Darwinism and to eugenics. "In the past the man has been first; in the future the system must be first," he stated. Consider what Taylor wrote regarding selecting manual workers for the job of handling a great weight of iron blocks per day:

one of the very first requirements for a man who is fit to handle pig iron as a regular occupation is that he shall be so stupid and so phlegmatic that he more nearly resembles in his mental make-up the ox than any other type. The man who is mentally alert and intelligent is for this very reason entirely unsuited to what would, for him, be the grinding monotony of work of this character.[5]

Henry Ford took Taylor's principles and applied them to the first mass production of cars on assembly lines. Čapek has the younger Rossum perform bioengineering that would make Taylor and the scientific management movement proud: "a working machine doesn't need to play the violin, it doesn't need to be happy, it doesn't need to do plenty of other things either." So he strips away all "tassels and artistic ornamentation" such as emotions because he wants a mass production line just like Ford: "manufacturing artificial workers is just like manufacturing diesel engines. The production ought to be as streamlined as possible, and the product ought to be perfect."

Yevgeny Zamyatin wrote his Russian dystopian novel *We* around 1919–1921, contemporaneously with *R.U.R.* (Interestingly, the second translation of *We* after the English one of 1924 seems to have been into Czech, published in 1927.) Zamyatin had experienced Taylorian mass production of ships in the UK during 1916–1917, spending two years supervising the building of icebreakers in the shipyards of Newcastle-upon-Tyne.[6] In his novel's totalitarian state people have become robot-like, with alphanumerical designations rather than names. The protagonist D-503 undergoes a forced operation to remove his imagination and emotions, an operation that turns people into "some kind of tractors in human form"; in other words, robots.

In addition to *R.U.R.* and *We*, we can see reactions to this efficiency movement in artworks such as Raoul Hausmann's 1921 *Spirit of Our Time*, and in films such as Fritz Lang's *Metropolis* (1927), adapted from his wife Thea von Harbou's 1925 novel, with its *Maschinenmensch* (machine person). Chaplin's 1936 film *Modern Times* has an automatic feeding machine for the factory workers tested on the Little Tramp; this Billows feeding machine is probably satirizing Charles Bedaux, another advocate of scientific management. Another famous scene from the film, the assembly line where the Little Tramp struggles to keep up with tightening bolts, is referring back to Taylor's observations of workers with a stopwatch, known as time and motion studies.

Čapek's Robots were assembled in a factory from synthetic biological materials. Domin and Helena's dialogue in the Prologue describes mixing the ingredients for Robots in a vat, containers for the preparation of livers and brains, a bone factory, and a spinning mill for weaving nerves and veins. But after the first performances of *R.U.R.*, robots escaped from Čapek's hands and immediately developed into electronic and mechanical machines. This change of imagined construction, from biological to electromechanical, perhaps came about because twentieth-century electronic and mechanical engineering were far more developed than contemporary biotechnology. Just prior to Čapek's *R.U.R.*, an earlier generation of remote-controlled autonomous machines included creations of Leonardo Torres Quevedo, the Telekino (1903) and the Ajedrecista (1914), as prominent examples;[7] as *Scientific American* put it in 1915, "He would substitute machinery for the human mind."[8]

Arturo Rosenblueth, Norbert Wiener, and Julian Bigelow wrote in 1943,

If an engineer were to design a robot, roughly similar in behavior to an animal organism, he would not attempt at present to make it out of proteins and other colloids. He would probably build it out of metallic parts, some dielectrics and many vacuum tubes. The movements of the robot could readily be much faster and more powerful than those of the original organism. Learning and memory, however, would be quite rudimentary. In future years, as the knowledge of colloids and proteins increases, future engineers may attempt the design of robots not only with a behavior, but also with a structure similar to that of a mammal. The ultimate model of a cat is of course another cat, whether it be born of still another cat or synthesized in a lab.[9]

The first such robots, autonomous machines thought of as artificial animals, appeared in the decades following *R.U.R.*, including Grey Walter's tortoise Machina speculatrix (1948), Norbert Wiener's moth Palomilla (1949), and Claude Shannon's mouse Theseus (1950). These creations—necessarily electronic rather than made from wetware—were at the beginnings of the fields of artificial intelligence and cybernetics. In 1948, Norbert Wiener formulated the principles of cybernetics, the basis of practical robotics.[10]

As robots turned from biology to mechanics, so too did other contemporary technologies. The first aircraft, for instance, began with more birdlike mechanisms—the Wright brothers' Flyer (1903) used wing warping to maneuver as birds do, not flaps and slats—but rapidly evolved into the complex mechanical machines that are airplanes today. Isaac Asimov, who coined the term robotics in his writing (in 1941), began to write his robot stories in 1939, a decade before Walter and two decades after Čapek. Asimov's positronic brains were thought of as an alloy of platinum and iridium, rather than grown from organic components. Despite being a biochemist, Asimov was far more interested in the software than in the hardware, and formulated his Three Laws of Robotics to explore the problems that could emerge from complex programming.[11]

Films, too, showed electromechanical rather than biological robots, perhaps because special effects on screen could much more easily portray a metal man—or woman, as in *Metropolis*—than one that is the result of biotechnology. More importantly, robots indistinguishable from humans would not serve the filmmaker's purpose; they must be made to look monstrous. H. G. Wells wrote a review of *Metropolis* for the *New York Times*, published in April 1927.

Possibly I dislike this soupy whirlpool none the less because I find decaying fragments of my own juvenile work of thirty years ago, *The Sleeper Awakes*, floating about in it.

Čapek's Robots have been lifted without apology, and that soulless mechanical monster of Mary Shelley's, who has fathered so many German inventions, breeds once more in this confusion.

Originality there is none. Independent thought, none.

. . . what this film anticipates is not unemployment, but drudge employment, which is precisely what is passing away. Its fabricators have not even realized that the machine ousts the drudge.[12]

Thus Wells dismissed the fear that mankind would be turned into robots. But Norbert Wiener commented, "The world of the future will be . . . not a comfortable hammock in which we can lie down to be waited upon by our robot slaves."[13]

It is curious that Wells describes Frankenstein's creature as mechanical. In fact, Čapek's Robots are more similar in their makeup to that distinguished ancestor. Mary Shelley's creation was, like them, biological, although Frankenstein directly reused body parts of dead humans for his creature rather than creating organs from protoplasm. This conception of the body as a machine can already be seen in Leonardo da Vinci's drawings, sketches, and automata from the fifteenth century.[14] Another creature, this time from the robot's home city, is the Golem, created from clay of the river bank of Prague; we shall return to clay later regarding the origin of life on Earth. The creation of the Golem is attributed to Rabbi Loew, a historical figure from late sixteenth-century Prague, but seems, like Frankenstein, in fact to be a creation of the early nineteenth century.[15]

Wiener's 1964 book *God & Golem, Inc.* concludes that "the machine . . . is the modern counterpart of the Golem of the Rabbi of Prague."[16] Čapek acknowledged that he based his Robots on the Golem together with the assembly line; "the Robot is the Golem made flesh by mass production," he wrote.[17] In a 1935 newspaper article, reprinted in translation as the afterword of this book, Čapek reacted strongly to the mechanization of his creation:

It appears, however, that today's world cares not for . . . scientific Robots and has replaced them with technological Robots; and these, undeniably, are the genuine flesh-of-our-flesh of our age. The world needed mechanical Robots, for it believes more in machines than in life; it is fascinated more by technological wonders than by the wonder of life.

Let us turn now to the wonder of life.

CHEMICAL GARDENS AND SYNTHETIC BIOLOGY

The story of chemical gardens begins in 1646 when Johann Glauber, alchemist, proto-chemist and proto-chemical engineer, published *Furni novi philosophici* (new philosophical furnaces), a textbook full of chemical knowledge. One of the chemical reactions he describes, the chemical garden, produced striking biological forms: "A water [i.e, solution] into which when any metal [i.e, metal salt] is put, it begins to grow within twenty-four hours time in the form of plants and trees, each metal according to its inmost colour and property, which metalline vegetations are called philosophical trees, both pleasant to the eye and of good use."[18] His textbook was widely read and spurred research of others, including Isaac Newton, into these phenomena. Newton, who spent much time on chemical research, wrote a manuscript *Of natures obvious laws & processes in vegetation*, probably in the first half of the 1670s, on his experiments on these chemical gardens, in which he writes of metal salts and "their vegetation in a glasse."[19]

By the mid nineteenth century, chemical gardens were being investigated for their relationship to biological cells and the shared phenomenon of osmosis. In Germany, Moritz Traube experimented on artificial cell building and osmosis,[20] and Wilhelm Pfeffer undertook detailed quantitative experiments on how osmosis functions.[21] As the nineteenth century passed into the twentieth, a scientific movement was in full swing. In Mexico, Alfonso Herrera was researching what he termed *plasmogenia*, or plasmogeny, "a new science of the origin of life."[22] In France, Stéphane Leduc was investigating *biologie synthétique*, synthetic biology, based on the same chemical-physical principles.[23] Working with chemical gardens, these researchers reproduced the appearance of biological forms including those of plants, fungi, and insects. They disagreed on some details, as researchers are wont to do, but agreed that their work would lead, as Leduc's 1911 book put it, to the "mechanism of life."

What influences might Čapek have received from early synthetic biology? Čapek has his scientist old Rossum begin work in 1920—"he undertook to replicate a living material, called protoplasm, by the process of chemical synthesis"—and in 1932 achieve just what some researchers in the field were attempting: "he discovered a matter that behaved exactly like living material, even though its chemical composition was markedly different." This new method by which life could be developed, "one that nature has never made use of," was "simpler, more malleable, and quicker," with "the whole tree of life growing out of it. . . . Ending with a human being! A human made of a different material than us."

Ideas from the field had been percolating outward from science into the public sphere since the latter half of the nineteenth century. An 1875 letter that Karl Marx sent to his friend Pyotr Lavrov gives one an idea of the intellectual diffusion of these concepts:

My dear Friend, When I visited you the day before yesterday I forgot to tell you an important piece of news of which you may not yet be aware. Traube, a Berlin physiologist, has succeeded in making artificial cells. Needless to say, they are not completely natural cells, being without a nucleus. If a colloidal solution, e.g. of gelatin, is combined with copper sulphate, etc., this produces globules surrounded by a membrane that can be made to grow by intussusception. Here, then, membrane formation and cell growth have left the realm of hypothesis! It marks a great step forward.[24]

In a 1911 article in *Scientific American*, we read that "the problem of artificial life is connected intimately with that of the origin of life, and also with that of the characteristics or distinctive properties of living matter," and that

the failure of scientists so far to produce "artificial life" is not to be charged against the science of biology. Very few of the attempts to produce "artificial life" have been made by biologists, who realize only too well the complexity of the problems involved. The biologists will be satisfied for a number of years to come if they succeed merely in analyzing what goes on in a living cell, in terms of physical and chemical processes. From time to time they will attempt to imitate a structure or a process by means of a working model; but they will not speak of artificial life until they are quite sure of all the conditions that play a part in this most intricate of phenomena.[25]

What eventually curtailed this line of research was the revelation through other biological researches of the immense complexity of even a single cell. It became clear that a simple physical-chemical system like a chemical garden could not develop into a living cell merely by mixing inorganic components with "protoplasm." By 1953, when Watson and Crick unveiled the double helix structure of DNA, the whole earlier field of synthetic biology had passed into history, so the term as it is now used is really a new foundation based on a science of gene manipulation that developed with the rise of the fields of biotechnology, genetic engineering, and molecular biology in the latter part of the twentieth century.

By the mid twentieth century, for scientists chemical gardens had become simply a common school chemistry demonstration, and a reaction to try with a chemistry set. The idea of chemical gardens continued to diffuse into human culture, however, as in the 1936 poem "Love Lies Sleeping," by Elizabeth Bishop, in which she notably uses the chemical garden analogy and expects that her reader should know what a chemical garden is.[26] Her plea for the modern city to be human is a response to the fears of mechanization that led Čapek to *R.U.R.* Three decades after *R.U.R.*, the German novelist Thomas Mann included chemical gardens in his 1947 novel *Dr. Faustus*.

I shall never forget the sight. The vessel of crystallization was three-quarters full of slightly muddy water—that is, dilute water-glass—and from the sandy bottom there strove upward a grotesque little landscape of variously coloured growths: a confused vegetation of blue, green, and brown shoots which reminded one of algae, mushrooms, attached polyps, also moss, then mussels, fruit pods, little trees or twigs from trees, here and there of limbs. It was the most remarkable sight I ever saw, and remarkable not so much for its appearance, strange and amazing though that was, as on account of its profoundly melancholy nature. For when Father Leverkühn asked us what we thought of it and we timidly answered him that they might be plants: "No," he replied, "they are not, they only act that way. But do not think the less of them. Precisely because they do, because they try to as hard as they can, they are worthy of all respect."[27]

The scientific neglect of chemical gardens began to change with work in the 1970s by David Double and colleagues,[28] and from that original trickle of research there is now a new scientific movement with a new name, chemobrionics, looking at chemical gardens as examples of self-organized precipitation membranes having semipermeable properties.[29]

ABIOGENESIS TO BIOLOGICAL ROBOTS

A fog of confusion held back origin-of-life research for a long time, developing around abiogenesis, the evolution of life or living organisms from inorganic or inanimate substances. The problem centers around the failure to distinguish between two phenomena organized around very different time scales: the development of life and the evolution of life. The confusion arrived with an incorrect interpretation given to the experiments of Louis Pasteur and others in the 1850s. Pasteur showed in 1859 that if one sterilized a broth, i.e., removed living matter, then "decay"—growth of microorganisms, fungi, flies, etc.—would not appear. This was correctly taken to argue against the spontaneous generation in a short span of time of an organism from inorganic components. It was then incorrectly extended as an argument against the evolution of life from nonlife over an immensely greater span of time.

In the first case we are thinking of how a single organism develops; a human from a fertilized egg in the womb, a process of some nine months, or a bacterium through fission, taking perhaps half an hour. The question has historically been whether such a process might occur spontaneously from inorganic matter—"spontaneous generation"—and whether it might be replicated by humankind as "artificial life." In the second case we are thinking, instead, about how the first proto-cells appeared, from which all modern life has evolved. Once one has a first type of cell, what is called the Last Universal Common Ancestor, Luca, of all life on Earth, Darwinian evolution can then act on it to produce other forms; as Charles Darwin put it, "having been originally breathed into a few forms or into one; and that, whilst this planet has gone cycling on according to the fixed law of gravity, from so simple a beginning endless forms most beautiful and most wonderful have been, and are being, evolved."[30]

The link between chemical gardens and life is now seen not in terms of the development of life through spontaneous generation, but instead in terms of the evolution of life until the moment Darwinian evolution can take over. This process, involving the self-organization of complex chemical reactions and the self-assembly of a proto-cell, surely took place over far greater time scales than the development of an organism from another, preexisting organism. We are probably concerned with a time scale of hundreds of thousands of years, if not more.

Where such a passage from nonlife to life took place on the early Earth is one of the great scientific questions of today. Darwin himself, although he would not venture an opinion in print, wrote in an 1871 letter to his friend Joseph Hooker that he envisaged a "warm little pond" as a possible environment.[31] Today that is still considered one option. Another possible environment in which life could have emerged was completely unknown to Darwin, however. Hydrothermal vents at the bottom of the oceans, where warm or hot water laden with minerals is pumped out into the seawater, depositing chimney-like structures that may reach many tens of meters in height, were only discovered in the 1970s. These hydrothermal vents are gigantic geological instances of chemical gardens. A particular type of hydrothermal vent that emits warm, alkaline water has been identified by Michael Russell as being of particular interest for the origin of life.[32] In their internal structure submarine alkaline hydrothermal vents show a fantastic array of tiny compartments separated by mineral membranes which would be ideal places within which complex chemistry could embark on the road to life.[33] Back in 1911, Leduc was prescient on this point:

The seas of the primary and secondary ages presented in a high degree the particular conditions favourable for the production of osmotic growths. During these long ages an exuberant growth of osmotic vegetation must have been produced in these primeval seas. All the substances which were capable of producing osmotic membranes by mutual contact sprang into growth—the soluble salts of calcium, carbonates, phosphates, silicates, albuminoid matter, became organized as osmotic productions—were born, developed, evolved, dissociated, and died. Millions of ephemeral forms must have succeeded one another in the natural evolution of that age, when the living world was represented by matter thus organized by osmosis.[34]

Prague's Golem was fashioned from clay, we are told, so it is noteworthy that origin-of-life studies returned in the twentieth century to the possible role of clay in the process. In 1951 J. D. Bernal suggested a key role for clay minerals.[35] Following this, in 1982 Graham Cairns-Smith argued that life on Earth evolved through natural selection from inorganic crystals and that clay minerals could store and replicate information and so act as "genetic candidates" or proto-genes. At the present time, the role of clay minerals in the origin of life at hydrothermal vents is being investigated.[36] Thus, from early twentieth-century synthetic biology we have come full circle and are now once again searching for an understanding of how chemical gardens are involved in the origin of life.[37]

Just as Ford famously said that one could have a car in any color as long as it was black, young Rossum thought that his artificial worker should be the cheapest whose requirements are the smallest. "The product of an engineer is technically at a higher pitch of perfection than a product of Nature," because "God hasn't the slightest notion of modern engineering." There is current research, the minimal genome project, to obtain the simplest living cell by taking an organism and stripping away genes until the minimal set for a functioning organism is found.[38] And alongside this top-down approach, there is another, bottom-up line of research to assemble a synthetic minimal cell.[39]

In the years following *R.U.R.*, robots became divorced from their biological conception. More recently, however, as biotechnology begins to get closer to Čapek's ideas, his biological Robots have been returning to fiction, but often under other names. In David Mitchell's novel 2004 novel *Cloud Atlas*, Sonmi-451 is a "fabricant," a clone made in a vat for slave labor and maintained in a submissive state through chemical manipulation of a foodstuff known as "soap." Those ideas return us to *R.U.R.* as well as to Taylor's ideal "stupid" laborer and to Zamyatin's D-503's forced operation: in *We*, D-503 lost his humanity; in *Cloud Atlas*, Sonmi-451 gains hers. Somewhat earlier, the 1982 film *Blade Runner* has "replicants," genetically engineered superhuman creatures. It is noteworthy that the novel the film was based on, Philip K. Dick's 1968 *Do Androids Dream of Electric Sheep?*, used the term android, and the film deliberately moved away from using that term. (In fact android, meaning a humanlike robot, is a word much older than robot. Ephraim Chambers's *Cyclopaedia* of 1728 contains perhaps the earliest use of the word, referring to a type of automaton.) *Star Trek*'s synthetic intelligent life form Data is likewise described as an android. A third related term is cyborg, "cybernetic organism," a 1960 coinage for a man-machine fusion. The 1970s television series about a "bionic man," *The Six Million Dollar Man*, and the 1984 film *The Terminator* employ this concept. All of these portrayals owe something to Čapek's biological Robots.

We are still far from Čapek's century-old vision of synthetic biology. It is just now, after the intervening century in which his Robots became associated with electronics and metal, that we are venturing along the path he envisioned of synthetic organisms fashioned from artificial biological materials. If we can fashion such artificial life, we should also be better placed to understand how life on Earth began from nonlife.

2

ANOTHER METHOD WITH THE POTENTIAL TO DEVELOP LIFE

Nathaniel Virgo

The Robots in *R.U.R.* are not mechanical devices but biological creatures. However, they are not "life as we know it," but something different. As old Rossum writes in his notes:

Nature discovered only one method of arranging living matter. There is, however, another, simpler, more malleable, and quicker method, one that nature has never made use of. This other method, which also has the potential to develop life, is the one I discovered today.

The story goes that old Rossum was researching protoplasm—the complex mixture of chemicals and enzymes that fills our cells, of which we'll speak more later—when he discovered "a matter that behaved exactly like living material, even though its chemical composition was markedly different." In the play, old Rossum is able to produce a doglike creature from this new substance within a few years and starts work on recreating humanity before his nephew, the young Rossum, takes over his research and uses it to create Robots instead. But is it plausible that we could ever discover such a thing—"another method" of arranging matter into life, different from the one we know today—and if so, what would it mean?

We know a lot more about the cell and its protoplasm today than we did in 1920 when *R.U.R.* was released. We know that at its core is DNA, a sort of molecular tape on which the instructions to build the cell are written. We also know that this tape can't read or reproduce itself. Instead, the protoplasm is a jiggling mass of complicated molecular machines called enzymes, each of which has its own special job to do. Some enzymes break down food and release its energy, while others use this energy to transport molecules from one part of the cell to another, or to build its membrane. Still others move along the tape of the DNA, either to copy it into another string of DNA or to make a temporary copy out of RNA. The RNA is then fed to yet another machine, one of the largest and most wondrous of all, known as the ribosome.

The ribosome reads the RNA three bases at a time. Through a fiendishly clever mechanism, it uses these three bases to select an amino acid. If it sees GGC it picks glycine, whereas CGA represents arginine, and so on. (The complete table it uses is known as the genetic code.) It then strings these amino acids together, one by one, in the order specified by the RNA, to build a protein. The protein then folds up in a way that depends on its sequence and starts acting as a molecular machine. It might be one of the enzymes that breaks down our food, or it might be part of the ribosome itself. The inside of the cell is made of machines that make machines that make machines. They use the DNA to store the information they need, but it's the machines—the proteins in the protoplasm—that do all the active work.

So that's the way the cell works. But how much of this has to be this way? This is a question that's really hard to answer, because although life on Earth has incredible diversity, if we take any organism and trace its genetic history back far enough, we always reach the same distant ancestor. All of us—humans, plants, bacteria—are part of the same giant family tree. Despite the diversity of life on Earth, it's really only one example of life.

But what if we found another? Another form of life that arose completely separately from the tree of life that we know? We might find this somewhere in our Solar system—on Mars, or Europa or Enceladus, or even in some extreme environment on Earth. Or we could spot it out among the stars, inferring its existence from measurements of the atmospheres of exoplanets. (Though in that case it would be a very long time before we could know anything about its biochemistry.)

Another possibility is that we could discover how to make it ourselves, just as old Rossum did. We could discover the right chemicals to mix together in the right environmental conditions, where molecular machines of some kind begin to form spontaneously, and start to extract energy from their surroundings and build other machines. They might sometimes build slightly different machines that build other, slightly different machines that might, by chance, perform better than their ancestors and be selected to form a new generation. If we found this, would we call it life? Perhaps, or perhaps not. But it would be something like a first step along the road to life, a potential ancestor of a whole new kind of life, even if it might take a million years to actually get there.

If we found one of these other forms of life, what would it be like? Although it's impossible to know for certain, I would guess that if we found another form of life in space or on Earth, we would find that it's made of molecular machines, and that it has some kind of "tape" like DNA, and that it would be mostly in the form of membrane-bound cells. But these machines would not be the same machines as in the life we know. They might or might not be proteins, but even if they are, they probably wouldn't use the same twenty amino acids that we use. The tape might not be DNA but some other molecule that can behave in similar ways. The larger machines, like the ribosome, would probably have discovered very different ways to do their jobs. Most of all, I would expect there to be huge surprises—things we never imagined could be done differently—because as the history of biology has shown, evolution rarely conforms to our expectations.

If we created our own new proto-life in the lab it might be even more different. It wouldn't have had millions of years of evolution with which to fine-tune its own mechanisms, and for this reason I doubt it would discover all the clever tricks like storing information on a tape or surrounding itself with a membrane to form cells, unless we put those features in by hand. But it would still tell us an enormous amount about the processes that lead to life, and it might even have practical uses: it would be made of enzyme-like machines that are different from the ones we know, and these might catalyze different kinds of chemical reaction. One can imagine possible applications in drug discovery, waste disposal, or materials science. It might just be a scientific curiosity, or it might open up a whole new kind of technology that's hard to imagine in advance.

Would we use it to build humanoid robots? Honestly, probably not—the brain is a complex thing, and having a new form of life wouldn't necessarily make it easier to build a new brain. But it would be a monumental discovery nonetheless, and perhaps it would change the world just as much as Rossum's work does in the play, but in different ways that are hard for us to imagine.

3

HUMANS AND MACHINES: DIFFERENCES AND SIMILARITIES

Carlos Gershenson

One of the most amazing things about reading *R.U.R.* a century after it was first published is noticing how many questions underlying the story are still current. Čapek's Robots are not mechanical but living. In this sense, they are closer to artificial life than to artificial intelligence. One should consider that the play was staged before the first electronic computers were built and before DNA was discovered (no mobile phones, no commercial aviation, no internet). We still do not have agreed definitions of life nor intelligence—imagine how ambiguous these should have been a century ago.

There is a thin line between technology leading to utopia and to dystopia, and science fiction tends to cross it constantly. On the one hand, technology could provide a eudaemonia, where everyone is happy and has their needs satisfied. On the other hand, something could "go wrong," leading to any type of postapocalyptic situation. Most probably, our future will be somewhere between those extremes. Many aspects have been improving due to our technology: life expectancy, literacy, economy, etc. As a species we've never had so much power and influence over our planet as we have today. However, this means our mistakes can have greater repercussions. Climate change is one example of this.

To understand better our future relationship with machines, it is useful to study the differences and similarities between humans and robots. Animals, children, and humans with different abilities can also be contrasted with Western stereotypes of adult humanness.

WHAT MAKES US HUMAN?

Comparing humans and robots is an interesting way of pondering what makes us human, a key element in *R.U.R.* There are several aspects to this question, and I focus on only five of them: soul, intelligence, understanding, generality, and emotions.

SOUL

In *R.U.R.*, Robots have no soul. What does that mean? It seems similar to the zombie problem related to consciousness: how do you distinguish a conscious human from a zombie that behaves exactly the same but has no consciousness? Another way of putting this: how do you prove that you are a conscious human and not a zombie? It would help if we had a proper definition of consciousness. We do not, nor can we properly define the soul. The word has different definitions in different contexts.

Perhaps the question should be: how sophisticated does a robot need to be to be considered human? If we do have a soul (whatever that is), and if robots become more and more like humans, we could assume that at some point they would also have a soul. Would the difference between soulless and soulful be qualitative or quantitative? Is it a gradual or a sharp transition?

Maybe our conceptual biases (on whether a robot is alive or not) might lead us far from the relevant aspects of these questions. So let us not speak about robots but animals. Do they have soul (or consciousness)? If they do, or if they have a "protosoul," how should we treat them? If robots had a soul or protosoul, how should we treat them?

To avoid potential spiritual or religious debates, some of these questions might be better posed if instead of soul we speak about mind, cognition, or intelligence.

INTELLIGENCE

Alan Turing proposed in 1950 his famous test to address the question "can machines think?" The test takes inspiration from the imitation game, where an interrogator tries to distinguish between a man and a woman by only asking questions. The woman tries to help the interrogator, while the man tries to deceive him. The woman wins if the interrogator identifies her correctly, the man wins if he is mistaken. Turing replaces the man by a machine: if it statistically manages to pass for a human, then it can be considered intelligent. Note that such a test is a central aspect in Philip K. Dick's 1968 novel *Do Androids Dream of Electric Sheep?* that inspired the 1982 film *Blade Runner*.

The Turing test was passed decades ago for nonexpert interrogators: computer programs introduced typing mistakes; naïvely, interrogators assumed that computers do not make such mistakes, so those who make them should be the humans. The test is passed, not by computers being as smart as humans, but by being as dumb as we are. In any case, even if a computer program were to pass the test, it would be only for conversational skills. We could define different Turing tests for different skills and argue that machines pass them once they become better at the tests than humans. Some have been passed already (arithmetic, chess, Jeopardy!, go, etc.), some might be passed soon, some might never be passed. For example, the goal of the international Robocup competition set in the 1990s was to have a team of soccer robots beating the world champions under FIFA rules by 2050. We've passed the halfway mark in time and the goal still seems very far away.

Many artificial intelligence systems are impressive, but they are still limited by context. They can beat humans, but only in the domain they were designed for. The so-called "artificial general intelligence" has so far produced only speculations.

Turing tests have been used with an inverted purpose: to distinguish humans from machines. In general, Turing tests are not perfect. I keep failing CAPTCHA tests (Completely Automated Public Turing test to tell Computers and Humans Apart). Does that mean that I am not human? Even were machines to pass Turing tests, perhaps we would still choose not to consider them intelligent.

Another problem with such tests is that not all humans may pass them. This might be because of particular abilities, specific impairments, or simply age. Many machines and animals will "beat" newborn humans. But I do not think that many people would care more for a nonhuman, no matter how well trained, expressive, and creative it is, than for a helpless human baby with a disability. It is simply not ethical. Perhaps our ethics will change as machines become more sophisticated and we better understand animals, but it is natural to be more empathic to creatures we feel closer to (family, friends, social circle, species, kingdom, etc.).

UNDERSTANDING

John Searle proposed in 1980 the Chinese room argument: imagine there is an English-speaking person in a room with a big book with instructions on how to manipulate Chinese symbols. They receive a sequence of such symbols, and then use the book to produce a message. The instructions are such that the output is correct. Like a computer, the person would pass a sort of Turing test, in the sense that people interacting with them assume they speak Chinese. However, they really do not understand anything.

We would have to assume that the instructions are able to contextualize the symbols, and from artificial intelligence research we can say that it is not easy, since we have not managed to do it (right, Siri?); but let us give Searle a chance. Still, we could apply the same Chinese room argument to ourselves: we can do many things, but we do not know how we do them (dance, love, hate, speak, think . . .). We can try to give explanations of everything we do, say, and think, but do we really understand any of this? Perhaps we are just zombies following instructions. And if our neurons do not understand anything, where does understanding come from? Perhaps it is just an epiphenomenon.

Take a dictionary in a language you do not speak. All words are defined in terms of other words of the same language. Eventually, some words will repeat. How do you ground their meaning? This is the "symbol-grounding problem," which still has no agreed solution. I am in favor of Wittgenstein's approach: the meaning is given by the use we make of symbols. Thus, if we make a "proper" use of a symbol, for all practical purposes, we can say that we understand the meaning of the symbol. (Of course, this "proper" can change depending on the context.) Therefore, we could say that if a machine makes "proper" use of symbols, then it understands them.

GENERALITY

If Robots in *R.U.R.* are simplified versions of humans, how could they be superior? This applies to any machine we have built so far. They can certainly be superior in some respects, but not in all human aspects. They might "win" in domain-specific tasks (particular Turing tests), but we have yet to see an "artificial general intelligence." We might never have one, as such a machine would need to be as complex as humans. But if that were the case, how different would they be from us? At the end, this also points to the question "what makes us human?"

Let us simplify the question: will we ever be able to build robots (biological or electronic) with abilities such as those from *R.U.R.*? Or are there some inherent limits? It depends. Some of our robots already have greater abilities than Čapek's Robots, and greater than humans. But some abilities seem to be more than a century away at the least. To have such robots, it seems that we would need first to build (wet) artificial life. This is because part of our intelligence depends on our biology, and the classic AI idea of separating the mind from the body—treating the brain as a computer—seems to be inadequate to reproduce humanlike behavior. So we would need to mimic not only human intelligence but also human bodies. And as to wet artificial life, protocells seem farther away than they did fifteen years ago.

And then we would need to develop multicellular artificial life much more complex than current xenobots. (See Josh Bongard's essay in this volume.) Genetic engineering might be an easier path, although this also seems farther away than it did around the turn of the century. But then, if robots were based on our own biology (as in *R.U.R.*), how different would they be from humans?

EMOTIONS

In *R.U.R.*, Robots have no emotions. But if they cannot laugh, how could they think, or even pass the Turing test? How separable are emotions from intelligence? Quoting Marvin Minsky: "The question is not whether intelligent machines can have any emotions, but whether machines can be intelligent without any emotions."[1]

Minsky defines emotions as behavior modulators: for the same "inputs," we behave differently depending on our emotions. Intelligent machines should show a similar dependence. In this way, emotions can be modeled as internal variables that codetermine "outputs."

One could argue that emotions make us do irrational things, but it is now accepted that rationality is only one part of intelligence. Moreover, we should not expect advanced artificial intelligence to be purely rational, simply because the underlying logics we use for modeling reasoning (and in general, for all formal systems) are limited. This is a consequence of Gödel's incompleteness theorems, and in a way it is also related to the symbol-grounding problem. The limits of pure rationality may be overcome through experience, pragmatically choosing what is (contextually) useful, rather than seeking absolute truths.

ROBOT ETHICS

We should consider two types of ethics: "how should robots treat us?" (*exrobot ethics*) and "how should we treat robots?" (*inrobot ethics*).

SLAVERY

In *R.U.R.*, there are several situations that make us reflect upon slavery. The development of human rights has considerably decreased the percentage of people in slavery (although due to the relatively recent population explosion, the total number of slaves worldwide is huge). The fight for animal rights also has been increasing. In *R.U.R.*, Helena Glory is the president of the League of Humanity that fights for robot rights. How different should these rights be? Should humans, animals, and robots have the same rights? If not, how to justify their differences?

There is still considerable human slavery, but it is socially disapproved in many cultures (although the destinies of some people, e.g., professional soccer players, are traded as if they were horses). Animal "slavery" is disapproved by some, as the growth of the vegan movement worldwide shows. But most approve robot "slavery." That is why we build them, their purpose, and in part they exist to free humans from hard labor. Still, a few centuries ago, people had a similar attitude toward human slaves. Slavery was their purpose, to free nobles from hard labor. This attitude has changed considerably, and is changing for animals, so perhaps it will change for robots as they become more like us. This might be one good reason not to make most robots more like humans, and even to highlight our differences. Slavery is harsh for the exploited, but profitable for the exploiters. If machines are much less complex than we are, it would be more difficult to judge that their work is harsh for them (what else would they do?), and we could profit from exploiting them. Certainly, the less complex machines are, the less things they will be able to do for us. On the one hand, specialization in machines can be positive: we can have several not-so-complex machines coordinating to perform several not-so-complex tasks, but overall they free us from complex labor. On the other hand, it is not so useful to build machines that are too similar too humans (as if we could): there are too many of us already!

MORAL MACHINES

The advent of autonomous vehicles has rekindled an interest in moral dilemmas such as the trolley problem, in which one should decide which human lives to save or sacrifice. There are several variations of the problem, e.g., involving where people will die and how the choice should be made of who should be saved. Slight nuances in how the problem is presented make humans change their choice. Will machines make the right choice? The safest option is not to leave it up to them, avoiding at all costs situations where machines might have to make any choice of this sort. If a machine performs an action with legal implications, such as killing someone, who is responsible? The designer? The owner? The victim? Different situations might lead to different verdicts, but in any case responsibility becomes diluted. For example, if a child or a pet damages another person's property, the parents/owners are responsible. But if a pet or a minor hurts someone, the pet might be killed, and the minor might go to a juvenile detention center (depending on the country). Decisions about "criminal" machines might also vary depending on the situation.

In any case, it might be useful if complex machines have some sort of "compassion." Asimov's three (or four) laws of robotics try to prevent machines from harming humans, but you can always find a loophole, and that is precisely what his stories exploit. There will be cases where machines hurt humans, especially when humans program machines to do so. But this will not be enough to stop "progress": we will build more machines and become more dependent on them.

PURPOSE: COMPETITION OR COOPERATION?

Machines *are* replacing workers, but we are not becoming superfluous. It's economics: we need consumers. Machines need us to buy them, otherwise nobody would make them. What would be their purpose?

In several sci-fi scenarios humans become expendable, but current tendencies in the real world are not pointing in any of those directions. We are evolving more toward a symbiosis than a conflict. Machines and humans complement each other: they free us from labor and extend our minds. We can do more than our Paleolithic and Neanderthal ancestors because of our technology, not because of our genes. If at some point we do not need to work, we will find some other purpose for our lives, as we have been doing since slavery was invented.

Still, also related to slavery, let us recall Hegel's writings about masters and slaves. If masters are idle and do not work, they lose contact with reality. Slaves transform the world, so they will become more capable than their masters. The same could happen with us. We are more and more dependent on our machines. But rather than becoming idle, it seems we are using them to increase our knowledge of and impact in the world. So perhaps the analogy of masters and slaves is not so appropriate for humans and machines. Just as it is problematic to refer to the brain as master and the body as slave, it seems more appropriate to speak about machines as part of us. We do take care of our body, but we cut our nails and hair without regret. Whether this growing partnership with machines will change the conception we have of "I" remains to be seen. Just as with the extended-mind theory, if our cognitive processes extend into the world, where does our mind end? If our labor extends beyond our body, where would it end? Probably it won't be too different from other tools we use: skates, skis, bicycles, hammers, knives, etc. can *feel* like extensions of our body, and as protheses become more sophisticated, the boundary of our body will become less distinct. But when we stop using these tools, our sense of self quickly adapts.

Some religious people might argue that it is sacrilege to pretend to improve what god made. But we have been doing so long before we even had gods. Our ability to change our environment defines us as humans. It is transforming the planet, and while many will suffer the upcoming changes, the most adaptable will benefit.

2120

A century ago, it would have been impossible to predict all the technologies that we have now. If technological evolution has been accelerating, how could we say anything about the world in 2120? There are so many options that even the most probable scenarios have a very low probability of occurring. It is easier to speak about things that will not happen, or that are highly unlikely to happen.

NO SINGULARITY

Some people believe in the so-called "singularity." Based on an accelerating technological growth, it is assumed that "soon" machines will be able to improve themselves faster than humans improve them, leading to a runaway reaction and deeply transforming our civilization.

The argument for the singularity is partly based on exponential accelerations such as Moore's law. However, Moore's law has already been decelerating (tendencies tend to change). Also, an exponential curve does not lead to a singularity—we would need a hyperbolic curve. Moreover, exponential Moore-like laws apply to hardware, not to software. Computers can be several orders of magnitude faster, but operating system startup times have been reduced only nominally.

Still, it could be argued that a quantitative change can lead to a qualitative change, as the capabilities of deep learning algorithms have shown only recently. But even when machine learning techniques are managing to solve new tasks constantly, they are still far from becoming as complex as humans (right, Alexa?).

ANOTHER AI WINTER?

There has been a growing expectation around AI in recent years, including some overpromising from the research community in order to get funding, repeating the experience of the two previous AI winters (1974–1980 and 1987–1993), where unmet expectations led to a decrease in funding and reduced interest in the field. We could be heading to another AI winter, but this possibility still seems far away. Several countries and companies have been investing and will be investing too much to afford an early disappointment. China announced they will dedicate $100 billion to AI over this decade, aiming to become a world leader by 2030. Probably they will reach that position earlier. However, the AI that is being developed is task-specific, not going in the direction of artificial general intelligence. This is good news for the AI industry, since it will be more difficult to disappoint, while current AI technology is opening profitable niches. This is also probably good news for humanity, as most doomsday scenarios require a generalist AI.

INHERENT LIMITS

So far, all our AI systems are based on formal logic. Almost a century ago, Gödel, Turing, and Chaitin proved the inherent limits of formal systems. Unless we find a new way of developing AI, it seems that these limits will prevail, independently of how much computational power increases. We must be able to program ambiguity, polysemy, homonymy, contextuality, sarcasm, etc. into computers, without losing the benefits of knowing that certain inputs will always produce the same output.

We know that the world cannot be fully described using only formal logic. But the question is whether we can find descriptions that are "good enough" that we can say that machines are "as good as" humans.

Somehow, we think. In theory, if we have a proper understanding of this process, we should be able to reproduce it in artifacts. But in practice, this understanding may well always be unreachable.

UTOPIA OR DYSTOPIA

It seems that we can dismiss technological apocalyptic futures, though it is useful to keep them in mind. And yet our technology seems far from liberating us. We still have the human factor.

Our technology has the potential to solve many of our problems, but technical solutions may create new social problems. For example, the cure for many diseases is available, but pharmaceutical companies try to maximize their profits in commercializing the cure, even if this implies that not everyone will be able to afford it. Since new technology will be commercialized, very probably it will not be accessible to all. So even when solutions to problems might be found, this does not imply that the market will allow these solutions to benefit everyone. In this case, technology could even increase the social gap, as only those with access to certain technologies will have certain problems solved. This is then no longer a technological problem but a social problem. But isn't this one of the things that makes us human?

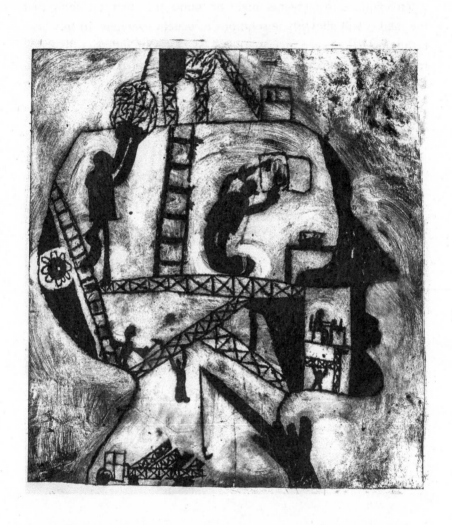

4

R.U.R. AND THE ROBOT REVOLUTION: INTELLIGENCE AND LABOR, SOCIETY AND AUTONOMY

Inman Harvey

God created man in his own image, so some would have us believe. Whatever view one has on that, undeniably a major feature of the contemporary world is the human endeavor to create machines in their own human image. In the context of this volume inspired by Čapek's play *R.U.R. (Rossum's Universal Robots)*, I focus on four aspects of this endeavor.

Any robot must have the ability to affect the world around it, and do so in some sensible nonrandom way. Even if these capacities are very limited, we can dignify them as examples of *intelligence* and *labor*. I class the design and achievement of these capacities to work intelligently as technical issues, matters for engineering solutions. Čapek asserts that his Robots have such capacities, but makes no useful observation about them. Most discussion in artificial life and artificial intelligence centers on these engineering problems, but I shall argue that these raise no difficulties in principle. Although there are plenty of fascinating and tough technical challenges that will keep people occupied for centuries, we have at least a rough picture of how to start tackling them.

But Čapek highlights two further aspects of robotics, issues of *society* and *autonomy*, that are more problematic and are not merely engineering issues. A core theme of the play is that the efficiency of Robots has a massive economic effect, displacing the livelihoods of human workers. Čapek's subtext was the politics of who benefits from the labor of others (whether human or robot). The autonomy of the labor force (again, whether human or robot) is mediated through social norms and constraints. Your freedom to do what you want lies within the possibilities offered by your social and economic environment—and needs to fit around my freedom to do what I want.

Where do autonomous robots fit into this? In the play, the autonomy of the Robots foundered on their inability to reproduce, or recreate further copies, in the absence of skilled human assistance. Autonomy (from the Greek for self-rule, having its own laws) can include an entity being in control of its needed resources (material needs, power resources); in charge of maintaining its own organization (self-repair to counter the ravages of entropy); in charge of its own acts. What *R.U.R.* brings to the fore is the interplay between society and autonomy. At the time it was written, the topic was science fiction. Now the robot revolution is turning into science fact.

PUBLIC PERCEPTIONS

Public perceptions of artificial life are dominated by two concepts: robots and computers. Both these terms were invented (in their current senses) within the last 100 years, and there are some surprising parallels in their origins.

The Czech word *robota* refers to forced human labor, and this was adapted by Čapek to refer to the mechanical servants of *R.U.R.* So the Robots were defined in terms of their subservient economic role alone, their humanlike characteristics going no further than the minimum needed for that role. Somewhat similarly, until the mid twentieth century the word "computer" referred to human office workers, perhaps in a bank or insurance company, occupied in humdrum lengthy and repetitive calculations. When a machine could take over such a role, the term computer carried over from the human worker to the machine. Again, it was only the functional role that mattered—in this case the ability to perform abstract calculations—and any further human characteristics were irrelevant.

Both robots and computers are thus originally defined in terms of severely limited subsets of human capacities, focusing on rather menial work and lacking in wider humanity; but this is sometimes forgotten. Young Rossum's requirements for a Robot led him to invent "a worker with the smallest number of needs. . . . He tossed out everything not directly related to the task at hand. . . . They are mechanically far superior to us, they have an astonishing capacity to reason intelligently, but they don't have a soul." A hundred years later some people, unaware of the history of these terms (which were deliberately limited in scope) but puzzled about how human nature relates to these technical artifacts, will perversely ask questions such as "are humans merely robots?" or "merely computers?" This makes no more sense than asking "is butter merely low-fat butter?" What these people probably really want to ask might be better phrased as: "are there any limitations to the range of human capacities that we can build into robots (or computers)?"

The idea of animate machines predates the terms robot and computer by millennia. More than 2,500 years ago Greek mythology tackled artificial life.[1] Homer talked of automata, and later Roman emperors such as Claudius enthused over their commissioned (and very real) automata, powered by pneumatics, used in theatrical spectacles.[2] "Automata" translates from the Greek for something that "acts by itself," implying something more self-willed than Rossum's original servant class of robot; when these fictional robots became masters of themselves, achieving self-willed status, it was the transition point at which they became a threat to humans.

Of course, Rossum's Robots were fictional; and the plot has them developing beyond their original limited roles, wanting to become masters rather than servants. In Japan—the country with the most robots— the more usual public perception is that robots are there to assist humans. By contrast, within European and Hollywood culture, the standard trope is for robots to be seen as an enemy threatening to take over humankind. There are at least two predominant versions of this threat.

The first version is that robots will take over all the jobs and leave a dispossessed class of humans; this has elements of truth, was a core concern of the R.U.R. play, and is discussed below. The second is that some singularity will be passed, when robot "intelligence" surpasses human "intelligence," and then the game will be over for humans. This I take to be absurd Hollywood fantasy, based on a confusion between reality and game shows and a naive understanding of what "intelligence" might be. There is no universal interspecies IQ test, and intelligence cannot be summarized in a single number. Rather, it is a somewhat fuzzy term referring to the degree of skill demonstrated by a person (and by extension, an animal or robot) at some set of tasks; as such it is a highly context-sensitive term.

INTELLIGENCE

In recent years artificial intelligence (AI) has progressed significantly, with a focus on the intelligence aspect of machines including robots, and specifically on abstract-reasoning intelligence. The straightforward computational side, the systematic performance of routine algorithms that 100 years ago was done by human "computers," has benefited enormously from hardware developments in speed, in the size of problems that can be tackled, and in the smallness and cheapness of the computing machinery. For some time there has been no contest on straightforward routine computational issues: computers beat humans, hands down.

The areas of intelligence that have presented more of a challenge have been the classes of formal problem solving that have resisted being reduced to algorithms. Many of these relate to pattern recognition, for instance identifying objects in images, or translating spoken Chinese to written English. For decades the orthodox GOFAI approaches—"good old-fashioned AI" that assumed all intelligence was at the core based on some algorithmic machinery equivalent to a Turing machine—made rather disappointing progress. But recent advances based on the competing methodology of neural networks have overtaken GOFAI approaches by leaps and bounds, pursuing a more intuitive concept of intelligence. Rebranded as deep learning—which largely means neural networks at scale, with big data sets—this has driven advances in speech recognition and image analysis that would have seemed incredible a decade ago but are now available to anyone with a smartphone.

Though currently much deep learning technology is implemented on top of a computational substrate, this is not at all necessary, and there may well be a significant shift away from this to different hardware (or wetware) substrates. The methodological move is away from programming a machine with explicit instructions on how to tackle a task, toward setting up the machine so that it learns for itself, from experience, how to tackle the task. AlphaZero, the system that within a few hours of self-play can master (to grand master level) sophisticated games such as chess, shogi, and go, would be the archetypal demonstration of this.[3]

Though a substantial amount of human programming went into jump-starting the ability to learn (and indeed to learn how to learn), by any plausible measure of effort the great majority of the skills developed are attributable to the nonprogrammed learning phase. In the longer term, the jump-start of human programming will fade into irrelevance, a historical detail. Of course, the defenders of GOFAI principles, who want to claim that all cognition is at root reducible to computation, will want to claim that even if a computational jump-start is a relatively small part of deep learning, it is nevertheless crucial and in some sense absolutely required. They are wrong when they make such arguments. That is not how natural intelligence in the biological world arose through Darwinian evolution, and evolutionary robotics demonstrates in principle how adaptive intelligent systems can be artificially evolved without the need for any such computational jump-start.[4]

Another argument put forward by the GOFAI advocates, unwilling to accept that they backed the wrong paradigm, is that the skills derived through deep learning are in some sense illegitimate, not real intelligence, because they are opaque. Nobody can explain just how the massively deep AlphaZero network actually ensures such novel and creative wins at go, or chess; it is suggested that this devalues the results. Such an argument fails to acknowledge that exactly the same shortcoming holds for the equivalent human skills. It does not need much reflection to realize that this is inevitable. Our understanding of other people, and of robots and machines, is necessarily limited in scope. Folk psychology gives us practical tools for interpersonal skills, psychology and cognitive science may take us a lot further, but any hope for a complete understanding of how humans work—or indeed of any robots that do a halfway decent job of emulating human performance—is clearly a fantasy. Whatever sense of the word is used, if "X understands X" means no more than "X can do what X can do" it is vacuous; but if it means more, then "X can do more than X can do" is a contradiction.

AI seems to me to have made the transition from being a field of over-ambitious promises with a poor delivery record to being a normal science that has become so essential to our everyday world that we take it for granted. There will always be the hype merchants, there will always be technical challenges yet to meet; but AI is now safe as an intellectual discipline. The computational GOFAI paradigm, though still fighting a rearguard action, is basically in retreat. We can move on from misguided attempts to see how intelligence can be built up from computations to more rewarding questions such as: How can our ability to compute be built up from our native noncomputational intelligence? How can natural disorganized material, such as "chemical soup," potentially form a substrate for computational systems?[5]

The methodological foundations of AI are now safe, but the dangers are elsewhere. We do indeed need to face up to the essential and inevitable opaqueness of robots when we consider regulation issues. The same sorts of issues arise in interpersonal regulation, in socialization of ethics and the functioning of legal controls. We return to this below when discussing society and autonomy.

LABOR

AI is largely focused on abstract intellectual problems, the kind that humans can do sitting in an armchair. Čapek was focused on robots that did useful labor; no sitting around in armchairs for them. Robots must be material, not abstract; must have physical engagement with the world. This physicality has tended to raise more tricky issues than the intelligence side of the equation.

Notoriously a robot vacuum cleaner is much more likely to be scuppered by the edge of a rug, or dog hairs trapped in the rollers, than it is by any abstract problem of planning a pathway across the room. Practical issues of power requirements and weight are still major problems.

Robotics has been bedeviled by its own version of the GOFAI syndrome, the assumption that any such physical problem can somehow be transformed into a computational problem. But there have been a variety of approaches to embodied robotics that reject this.

Common themes include the realizations that no design can be validated in simulation alone—only tests in the real world count—and that action in the world involves dynamics. Getting robots to walk or ride a bicycle is not a sequence of static issues to be solved but a question of matching the dynamics of forces changing in real time with the effects on the robot-world interfaces; and this has to be reflected in the dynamics of control systems, of "brains." Even though the neural networks of AI are often geared toward static problems, there is a whole field, as yet rather underexploited, of dynamical neural networks.[6]

Robotics has not yet reached the levels of achievement shown in AI: it has been the more backward sibling. The scope and range of possible physical interactions is far wider than for abstract intelligence. Nevertheless, I currently see no obvious barrier in principle to developing robots to achieve any task humans or animals can do. The current apparently insuperable difficulty is with emulating the concomitant ability of biological systems to be self-creating and self-maintaining; we return to this below in the section on autonomy.

SOCIETY

The master-servant relationship is a social one, involving economics and politics. *R.U.R.* is a political play and focuses on these issues more than the scientific ones for robotics. Economic concerns are paramount. The Robots are provided with pain—but only so as to protect them against damaging themselves. "Why don't you create a soul for them? . . . That's not in our power. . . . It'd increase production costs."

The efficiency benefits for their human masters are initially viewed optimistically. "In five years the cost of everything will be zero point nothing. . . . There's no more poverty. Yes, the workers will lose their jobs. But by then there will be no more jobs. Everything will be produced by living machines, and humans will only do things they love doing. . . . They'll live only in order to better themselves." Of course, this promise of Paradise does not come to fruition. After a few years "the workers rose up against the Robots and smashed them to pieces . . . various governments created Robot armies." The working people turned out not to share in the economic benefits, and those in power attempted to exploit them so as to maintain their power.

This prescient play from a hundred years ago echoes the sorts of explanations often offered for contemporary political shifts such as Trump's America and Brexit in the UK. Change can leave many dispossessed and disadvantaged; the reactions may be expected to be sometimes ugly. This is the real threat of the robot revolution.

Robert Shiller[7] catalogs numerous historical instances of popular narratives decrying technological developments that threatened to cause unemployment. Ned Ludd was (perhaps) a weaver who in 1779 smashed the knitting frames whose efficiencies had the side effect of putting weavers out of jobs. Whether a real or mythical event, the story led to the nineteenth-century Luddite movement taking his name. Subsequent depressions through to the present have often been attributed in part to labor-saving technology. Clearly the invention of the automobile was going to make redundant a lot of ironmongers making horseshoes; were the new jobs in car mechanics and industry sufficient to compensate?

Shiller (who cites Čapek and R.U.R.) cautions that we should acknowledge that such economic narratives may have an influential life of their own, regardless of how factually based they may be. But it is indisputable that automation and technology have already had enormous effects on employment; consider the agricultural revolution that reduces the proportion of people working on the land from greater than 70 to less than 5 percent. In the form of the robot revolution, in both AI and robotics, this can only increase in scope and in speed of change. Unskilled and semiskilled jobs are the first to transition—like those of the human computers who used to work in banking and insurance offices.

Car and truck driving still present challenges, but will eventually be automated. The professions that rely on human judgment (another name for intuition) will fall to advances in AI; radiology and other medical image classification soon, language translation and basic legal practice work soon to follow.

The automation of jobs may have unexpected side effects. For instance, finding a parking space in town is no problem if your car can just cruise around aimlessly while you shop or lunch—but your personal convenience is at what cost in traffic congestion for others? However, it is entirely predictable that any new jobs replacing those taken over will be more skilled rather than less, and in the transition a lot of people will be disadvantaged. Of course, in an ideal world, the optimistic one at the start of R.U.R., the economic benefits of automation could in principle be widely shared; but we know this does not happen inevitably and by magic. The internet, the World Wide Web, promised that when the circulation of information became nearly free, the result would be democratization of such resources. Such good effects have indeed largely happened across the world—but with the unanticipated side effects of corporate control of social media platforms, and the balkanization of political commentary.

Robotics and automation do indeed present very real dangers to mankind. But these are not technological dangers, nor some singularity fantasy of *Terminator* robots seizing control. They are human dangers. If robot missiles are given autonomous powers to fire at will, it is because human governments have so chosen. If economic changes arising from automation result in people being dispossessed, it is because human society has failed to take care of its own. As with climate change, robot change requires political solutions; it depends on who has the power to effect social changes in response to natural events. Both human exploitation of fossil fuels and human ingenuity in fashioning technology are natural events, but we need to face up to the human consequences.

AUTONOMY

Robots should not need constant monitoring and supervision by humans, and in that sense some degree of autonomy is generally essential to their purpose. How much? A robot (or a person) sent on a space mission to Mars may have no say in what their designated tasks are, yet necessarily must be given some independence, some autonomy, in just how these tasks are to be carried out; after all, it may take tens of minutes to consult by radio with Earth while a response is needed immediately. Some sliding scale of increasing autonomy would seem to go with more independence of choice as to what one chooses to do. But this is within a context of what opportunities and support there are from the surrounding environment.

Beyond the rather extreme example of Robinson Crusoe on a desert island, human autonomy relies upon support from the social and economic systems around one. Without ready access to food in exchange for money, it would fairly quickly become obvious to me just how limited my own autonomy is. Nevertheless, within such interplay between personal freedom and social and environmental constraints there is some scope for autonomy—in principle as much for robots as there is for humans.

Some people think that autonomy is a binary, all-or-nothing property, and that there is some objection in principle to robots having it. The view

I take is that it is a relatively innocuous and context-sensitive description that raises no deep philosophical issues. A central heating thermostat is autonomous insofar as it switches the boiler on and off in response to room temperature—but typically not autonomous insofar as somebody else designed it and set it in the room attuned to a specific room temperature. An artificially evolved robot control system[8] is autonomous insofar as it is "responsible for its own actions."

Within the framework of robots interacting with humans, with varying levels of responsibility for their own actions, there is a vast range of possibilities. At one extreme, hitchBOT[9] relied on appealing to the good nature of drivers in hitchhiking 10,000 kilometers across Canada in 2014, keeping a record of its adventures. Similarly reliant on human motive power was Norman White's "Helpless Robot," exhibited at the 1997 European Conference on Artificial Life. Positioned in a gallery, when it senses passing visitors it plaintively appeals to be given a push to rotate one way or the other. Once a human-robot dialogue has developed (pushes from the one, requests from the other), the robot becomes increasingly demanding, aggressive even, and then grumpy when abandoned.[10]

These playful examples illustrate that human-robot interaction can take many forms, and even a small degree of autonomy can (literally) travel a long way. The autonomy or freedom to act of a robot (or other human) is only a threat insofar as its exercise diminishes the autonomy of others. This is where social norms and regulations, and their enforcement, become relevant. Undoubtedly these will have to be extended to accommodate robotic advances, with regulation for autonomous driving an obvious immediate concern.

In R.U.R., the area where the Robots' lack of autonomy crucially let them down was their inability—without assistance from knowledgeable and skilled humans—to repair and replace themselves as they degraded and wore out. Presciently, this relates to what may well still be the biggest unsolved issue in artificial life: biological systems are self-creating, self-repairing; can we design material synthetic systems to be the same? We have theoretical approaches, such as autopoiesis, that offer possible routes to understanding what is necessary. In natural evolution there likely was a genuine singularity, when the first autopoietic entity arose that had potential for evolution—and thus instigated the living world we see around us now. But as yet there is no consensus on how to even start to synthesize the equivalent.

Some people will suggest that a similar "we haven't a clue" issue is that of providing a robot with consciousness. Does it make sense to talk of "inflicting pain" on a robot, over and above "inflicting damage," and should this feed through to how we regulate human-robot interaction? One can steal one's neighbor's horse and also inflict suffering on it, two crimes; one can only steal his cabbage. But how about his robot? The consciousness issue is to my mind a philosophical nonissue.[11] It arises from linguistic confusion, from treating subjective consciousness as an objective property that a robot may or may not "have." But "subjective" simply cannot be treated as if objective; that misunderstands the way such concepts work. On the "can robots feel pain" issue I am tempted to follow the same strategy that Turing used on "can machines think": "Nevertheless I believe that at the end of the century the use of words and general educated opinion will have altered so much that one will be able to speak of machines thinking without expecting to be contradicted."[12]

Turing used this in the context of proposing the Turing test, a public comparison between the (anonymized) performance of a computer and a person. His guess as to how far computers would go, by the end of the twentieth century, turned out to be at least roughly in the right ballpark. But his prediction is ultimately phrased in terms of how "the use of words and general educated opinion" will alter. My prediction would be that robot developments for social interaction, both with other robots and with humans, will be aided by visible/audible expressions associated with both physical damage and thwarting of intention, and that taking these as expressions of "pain" and "irritation" will become commonplace and accepted. After some period (Turing's fifty years seems again as good a guess as any) the use of words and general educated opinion will feel comfortable dropping the scare quotes around "pain" and "irritation."

SUMMARY

Those who worry about some terrifying robot apocalypse, triggered by some singularity, will no doubt think the comments above miss the main point. But I consider such fantasies absurd, and as diverting attention from the very real dangers arising from accelerating advances in robotics.

There are fascinating technical challenges in developing the intelligence and physical capabilities of robots, challenges that will no doubt keep people busy for decades and centuries to come. But we know enough to know how to continue tackling them. There are no obvious difficulties in principle now that the computational GOFAI logjam has been convincingly broken.

For decades AI was held back by the confusion between computing as one skill that people could perform, and computing as the mechanism by which this was done. The triumphs of deep learning are not just important as a marker for how we can start to mechanize intuition and learning as well as reasoning, but also as a step toward understanding the relationship between mechanism and performance. This lesson extends well beyond the neural network mechanisms currently used.

What we should be worried about, above all else, is the social, economic, and political impact arising from robotics and automation. The robotic revolution will be at least as widespread as the agricultural and industrial revolutions, and may well happen much faster. Much economic benefit will result, but no doubt this will, if unregulated, be shared unfairly. In consequent social and economic upheavals there will likely be large groups of people made redundant and unable to take full advantage of the new opportunities available; they will be left behind.

Robots taking over human jobs will inevitably have varying degrees of autonomy, and interact with humans in somewhat humanlike fashion. Hence both sides in these interactions will face the normal human issues of when my autonomy interferes with yours. Social norms and regulations will have to be tailored to fit new circumstances. But social norms do not just ease everyday interactions; at a deeper level they also delineate the framework of political power.

A hundred years ago Čapek presciently anticipated this social and economic upheaval in his play *R.U.R.* The coming robot revolution is inevitable. We will need to incorporate the new robot interactions into the social contract(s) of everyday life and, crucially, address the political issues of how economic benefits and power and control are distributed in the aftermath of this revolution.

5

IT WASN'T WRONG TO DREAM: THE PARADISE OR HELL OF OUR JOBLESS FUTURE

Julie Nováková

It wasn't wrong to dream of forever eliminating the drudgery of work; the humiliating and terrible burden of human labor that enslaved humanity, and the dirty and murderous daily toil. Oh, Alquist, it was too hard to work. It was too hard to live.

Čapek's timeless drama asks the age-old question of what it means to be human, and offers prescient visions such as artificial life itself (both from the perspective of robot workers and the vision of synthetic biology) or the inner workings of an artificial mind. But there is another theme, less conspicuous in the work but increasingly relevant in the current world and closely tied to the robotization of the work force: the loss of demand for human work.

This is by no means a new threat. The early nineteenth century saw the Luddite movement of textile workers opposed to the use of machinery in the industry. The workers feared that with the rise of machines, they and their hard-earned skills would become obsolete.

The following two centuries showed that their concern was largely unfounded. Machines replaced some human jobs, but created other ones. The composition of the work force changed radically, but machines brought forth no radical shortage of jobs.

That, however, might finally change with the rise of robots and AI. In a famous 2017 analysis working with parameters such as social intelligence, creativity, and manipulation in a given job, Oxford University researchers Carl Benedict and Michael A. Osborne concluded that nearly half of all employment in the United States, generalizable to most of the developed world, is at high risk of replacement by automation, most importantly office, service, and sales work.[1] Some towns already face the problem of mass unemployment.[2]

There is one widely discussed proposed solution: universal basic income (UBI), an unconditional monetary allotment for every citizen to cover their basic needs—shelter, food, health. Imagine a future where you don't need to work for a living. You can devote your time to your true passions instead of toiling in a workshop or an office. You could pursue learning, read as much as you'd like.

Or not. Čapek saw the notions of humankind freed from work as entirely negative. His Harry Domin, general manager of Rossum's Universal Robots, lays out his grand, proud vision early in the Prologue: "in less than ten years Rossum's Universal Robots will be producing so much wheat, so much material, so much of everything, that we'll say: things have lost their value, everything is free. . . . There's no more poverty. . . . Everything will be produced by living machines. . . . Work will disappear. Humans will only do things they love doing, and they'll be freed from all care, and liberated from the indignity of labor."

It sounds wonderful, doesn't it? But in Čapek's imagination, it throws humanity into disaster. Without the need to work, people become lazy and dependent on Robots. Humans are no longer needed; children are no longer being born; all the while wars caused by the abundance of Robot soldiers plague the Earth.

While we can dismiss the lack of births as a metaphor rather than a prediction, the question remains: Will humans become lazy and reliant on outside assistance once most work becomes unnecessary? Experimental and theoretical works continuously aim to constrain the range of probable outcomes, but there is one more way to explore the impacts of UBI in the long term: science fiction.

The genre acts as a sandbox, exploring ideas from Čapek's Robots and their fallout to what could happen if humans could change ourselves immensely, be it by bioengineering, cyborgization, tailor-made drugs, or other means. Perhaps it's time science fiction started focusing more on the future of work and possible options and consequences of UBI or other potential solutions. Recent novels such as William Gibson's *The Peripheral* or Annalee Newitz's *Autonomy* depict the advantages and pitfalls of societies that are not post-scarcity, but where work has become another scarce resource—in other words, the likely near future. *The Absolved* by Matthew Binder paints a very dystopian picture of UBI, adhering to the view that it would essentially make people lazy and unhappy.

Other works of science fiction depict a far more optimistic post-scarcity future. Take *Star Trek*, for example. Money is no longer used for everyday transactions in the United Federation of Planets; it would be more precise to say that such transactions no longer exist. Everyone has shelter; everyone can synthesize food, clothes, and other items of basic needs from the ubiquitous replicators. Yet society doesn't stagnate; at least we're not shown that side of it. Instead, we witness men and women boldly going where no one has gone before, be it in the actual starships or in the more abstract realms of science and art.

Other fictional creations are more ambiguous in their portrayal of the impact of UBI or UBI-like prospects. The TV series *The Expanse*, based on the novels of James S. Corey, portrays an overpopulated Earth where work is the scarce resource—work and any luxuries. Though everyone is entitled to goods and services that should suffice for basic subsistence, in practice those without other preexisting means are forced to live in cramped conditions, only receive the basic foodstuffs and medicine, have no means to travel, and have practically no chance to land a job however much they want to—in this setting, the "Basic" (the goods and services allowance) and employment are mutually exclusive, and while most people wish for the latter, they have to settle for the former. Space settlements provide a stark contrast to this world that Martians and Belters consider a "degenerate paradise" of sorts; everyone has to work "out there" in order to survive. The Martians strive for the distant shared goal of terraforming the barren planet, which everyone has to be committed to, while the Belters are the work force trodden on both by the Earth and Mars, surviving in their ships and asteroid habitats, mining rare metals and ice. Once outside the plush but crowded conditions of the home world, humans have to work hard; on Earth, they mostly cannot work even if they want to. *The Expanse* illustrates the possibility that poverty won't be eliminated by giving people what they need to survive—it will just change. While poverty has become non-life-threatening on Earth in *The Expanse*, so many of the "useless" things that make us human, as Harry Domin notes in the Prologue of *R.U.R.*, remain out of reach for the vast majority of people.

There is an important distinction to note: *The Expanse*'s "Basic" provides goods and services, not money. Author Scott Santens argues that such is one of the worst ways to try to eliminate poverty, and should not be confused with UBI. In summary, "Basic income is a floor. The Expanse's Basic is a ceiling. . . . Basic income is freedom. *The Expanse*'s Basic is control."[3]

Some people still envision a future with provisions like that of *The Expanse*, although small-scale experiments, as we'll see in a moment, have mostly painted a very different picture. Could such a dystopian future still happen with monetary UBI, though? While not everyone can become an artist, a scientist, or a businessman; while not everyone can achieve high formal education; while not everyone is curious, are we still slated to become a cramped-quarters-confined, Netflix-consuming, lazy bunch—even if we most certainly don't want to?

Only practice can show what happens. No large-scale tests have been run so far, but smaller-scale experiments since the 1960s mostly show improvement. They are already far too numerous to list them all here, but we can extrapolate that UBI can lead to reduction of health problems (especially in mental health and stress-related issues), better school results and attendance of children, better parent-children relationships, increase in savings, elimination of starvation, better protection of the environment, and an increase of new business enterprises with the potential to create new jobs. An incomplete but growing list of studies on UBI and UBI-like trials can be found in the Basic Income Earth Network's research depository;[4] a detailed summary exists in the Basic Income Reddit community.[5] Apart from special demographics such as new mothers and teenagers, UBI didn't seem to make people any less (or more) likely to find a job. Moreover, UBI can support entrepreneurship by providing a floor so that people are more willing to take risks without the threat of descending into poverty.[6] Finally, a recent meta-study showed that worries about people spending their money in an "easy-come, easy-go" way on alcohol, tobacco, and such were largely unwarranted.[7]

The impacts from these small-scale experiments seem to be overwhelmingly positive. However, the question remains of whether UBI can be economically sustainable. Plunging a country deep into growing debt is not a long-term solution; UBI implementation should curb much of government bureaucracy and save money spent on targeted poverty relief, medical costs, policing, and more. But for long-term sustainability or even profitability, only untested predictions and calculations exist so far—though many seem optimistic (at least for the developed world), such as a recent estimate by Karl Widerquist.[8]

UBI implementation is currently on the agenda of political parties and other organizations across the political spectrum and in many countries. More or less detailed and serious-minded proposals exist for Italy, Finland, Switzerland, Slovenia, Australia, the United States, and other states. Perhaps, then, in our lifetime we will get to see the actual impacts of nationwide UBI—and with it hopefully see the elimination of poverty and growth of work on people's own terms.

In Act II of *R.U.R.*, general manager Harry Domin says: "I wanted people to become masters of their own destiny, to live for more than just a piece of bread! I wanted a world where no human mind is wasted on stupefying labor at somebody else's machine, a world where nothing, but absolutely nothing, is left over from that damned mess of the existing social order!"

While we may not think highly of Domin's arrogance and myopia, most of us would likely agree that one shouldn't live merely for the crust of bread and that raising people from poverty is good. Robotization of work, if sustainable and accompanied by appropriate political, social, and economic measures, can perhaps help us achieve that noble goal. Only further trials will reveal if that is truly so, and if it can also uplift humankind in the way Domin liked to envision.

6

R.U.R.: A SHREWD PLUTOCRAT, A GENIUS ENGINEER, AND AN ANTI-SUE WALK INTO A BAR

Lana Sinapayen

The cover of my edition of *Rossum's Universal Robots* features a man with stylized electronic circuits on his skin.[1] While electronics are the first thing that comes to the mind of contemporary readers when imagining robots, Čapek's story does not mention electronics; it was written at a time before computers, and before algorithms became the metaphor of choice to talk about intelligence and the mind. No metaphors in *R.U.R.*: the Robots are humanoids of flesh and bones. But not our flesh and bones; they are similar to us, yet eerily different. Čapek's book left me with a similarly eerie feeling: his preoccupations as a writer, the social issues influencing his play, and his blind spots as an observer of human nature are similar to those of many modern writings about robots, yet unmistakably different. The basic science of *R.U.R.* is at odds with the philosophy of current artificial life research. From these differences, we can imagine possible consequences for the Robots in the play, and infer some of the real-life biases that influenced the way Čapek wrote his characters. This essay ends by drawing parallels between the psychology of Domin and some of the most egregious crises of our time, a testament to the timelessness of Čapek's social criticism.

THE SCIENCE OF *R.U.R.*

For an artificial life researcher, it is particularly interesting to try to give scientific grounds to the "living matter" used by old and young Rossum to make their Robots. This matter results in the same biological laws as we know, yet it is chemically "markedly different": the elusive second-ever discovered example of life, without having to leave the solar system, "simpler, more malleable, and quicker" than our good old biology!

Reading between the lines, this material sounds like both a dream and a nightmare. The nightmarish aspect is the apparent absence of an immune system in these mysterious clumps of cells, that have such an "insane will to live" that Rossum could "stitch and mix it any way he wanted." This is a recipe for a catastrophe! Wouldn't the cells fuse to any organic matter that they make contact with? I like to imagine that this was one of the reasons old Rossum had such a hard time building viable bodies: organs fusing together, bone cells turning into muscle, eyelids melting into each other . . . these are consequences difficult to avoid when your base material is some kind of very promiscuous stem cell. And don't you dare let your creations shake hands.

The promiscuity of the base material would also be a good explanation for why this different kind of life stayed hidden and undiscovered for so long: it is just not a very successful model. The biology that we know is made of stuff that does not mix easily, which is why you can have different types of cells in a body, different types of individuals in a species, and different species in an environment. If organisms fuse too readily, you lose the possibility for differentiation, and you cannot access what makes evolution such a prolific source of creativity and robustness: diversity.

On the other hand, the material has obvious advantages: it can be used to make any kind of tissue from bones to eyes, in any shape. Its usage seems to be limited only by the imagination of the user. And yet, as Domin laments, the Rossums uncle and nephew chose to stick to making boring, boring humans. Why not self-repairing boats and photosynthetic buildings, bioluminescent submarines or flying musical instruments? If humanity is to be destroyed for playing god with the ultimate stem cells, at least let us go in style!

What limited the engineering pair could have been their insistence in building organisms in a top-down fashion. Their creations are limited to their own imaginations, unlike what happens with a bottom-up random search process like evolution. Evolution covers a lot of ground in the space of what is possible, as long as it works. As Rossum junior points out, evolution also has its disadvantages, notably its dependence on history. Evolution is a suboptimal optimization process: organisms are made of bits and bobs that happened to get the job done at the time when they appeared. Some of these characteristics are likely to be "useless," as Rossum assumed. So indeed, why not replace evolution by pure engineering?

ROSSUM AND THE COST OF EMOTIONS

It is not necessary to conceive of this machine as having any vegetative or sensitive soul or any other principle of movement and life, apart from blood and its spirits, which are agitated by the heat of the fire burning continuously in the heart.

René Descartes, *Treatise on Man*

Unfortunately, the cost function that Rossum chose in his optimization process was, literally, a "cost" function. The best worker is the cheapest, says Domin. Alas, as the main characters discover too late in the story, you cannot decide what is useless just by looking at a body and hacking out parts you do not like. There are just too many parameters to take into consideration: what time scale you are considering, what task you are optimizing for, what kind of environment your organism will live in, and so on. Ironically, if Rossum had used some kind of evolutionary process, he would have discovered that his creations more often than not would have ended up with emotions. Of course, he could also just have looked outside the window and noticed that all vertebrates have emotions, which they express in a way clear enough for us humans to empathize with. My pet theory is that Rossum did know all of this, and that was exactly what bothered him.

Emotions are a crucial tool for evaluating situations, taking split-second decisions, learning, communicating, memorizing, and bonding with others. Since Rossum's Robots had to learn how to do their jobs, and were very successful at doing so, I doubt that they were really built "without passion." Instead, they must have had a way to process verbal praise and scolding, a way to understand the emotions of their owners, a way to prioritize important information against irrelevant details, a way to direct and redirect their attention in the blink of an eye, a way to filter which memories to store and which to forget. Rossum might not have wanted to call it "passion" or emotion, but that regulation, alert, and filtering system must have performed very much like regular human emotions. Having managed to build such humanlike robots, the engineer could not have overlooked the importance of emotions for everyday tasks. What if he had instead wanted to prevent the expression of these emotions through body language, facial expressions, and voice? Rossum would then have needed a function acting like a stop valve. Could the "gnashing of teeth" and rage outbursts of early Robots have been symptoms of the breaking of the valve? An additional clue can be found in Dr. Gall's difficulty in "giving" emotions to Robots: his unnecessary reimplementation of an emotional system might still have been stumped by Rossum's secret backstop.

Why would Rossum go so far to hide something as basic as emotions? After all, he did give his Robots something he considered useless: gender. There are male and female Robots, because "people are used to" having a female typist and other gendered jobs. So why conceal emotions, which are far more useful? Miss Glory might have offered us a hint, by explicitly associating feelings with the idea of having a "soul." Denying a soul to an organism is an ancient trick to justify brutal exploitation: in different places and at different times, women were denied a soul, Black people were denied a soul, children were denied a soul, and of course nonhuman animals are denied a soul too. Humans are very good at reading or imagining emotions on the faces of others. Robots are things, goods, "spoiled freight." A Robot with visible emotions might have been too big a risk for the economically minded Rossum.

GLORIA PATRI . . . ARCHY

Gloria Patri, et Filio, et Spiritui Sancto, sicut erat in principio, et nunc, et semper, et in saecula saeculorum. Amen.

If digging into the science of *R.U.R.* and dreaming up backstories is a pleasant endeavor, the narrative tropes themselves are quite grating. Specifically, the character of Helena, the only notable female character, is presented as the archetype of the insufferably dumb yet treacherous woman; the only thing worse than the way she acts is the way she is treated by Domin. Even the Robot modeled after Miss Glory, Helena, is in the words of Dr. Gall "delightful and a complete airhead."

Helena lies her way into the Robot factory before revealing her true intent: liberating the Robots! For you see, she is a member of the League of Humanity, a group that cares so deeply about the well-being of their fellow humans. Inexplicably, while the League cares very much for Robots, the poor of Europe do not seem to count as "fellow humans." Glory does not know or care about the price of bread, fabric, or the dangers of factory work. If the League is not very concerned with the plight of its fellow white people, it cares even less about the lives of people of color, notably Native Americans, who according to Alquist himself do not live in enviable conditions. Despite the obvious parallels between Robots and enslaved Africans, Helena never mentions slavery, so we can guess she does not care about that either. She is the embodiment of the "brainless femme fatale" trope: fully driven by her emotions, incredibly shallow, and deadly. The polar opposite of the blank-faced, highly intelligent Robots!

Domin quickly talks her out of her humanistic feelings, force-kisses her, and despite Helena never saying "yes," they are thus instantly engaged. Hurray! The other characters immediately congratulate her for marrying a forceful man who cares not for what she says, as he spends most of the Prologue interrupting her; nor for her feelings, which are irrelevant to his physical attraction and the strings of his wallet. Later in the story, she dooms both humans and Robots to destruction twice over. First, she pushes Dr. Gall to modify the Robots, and eventually burns the precious manuscript that could have elevated the Robots to the status of reproducing species, and could have been used to bargain the lives of the last handful of humans on Earth. What a pity, Helena, that you had to be written as a brainless traitor of a woman, a lowly form of life governed solely by emotions, whose existence serves only to make men lose their heads.

Helena is a plot device more than a character; her function is to throw the story into chaos without the author having to justify her actions or the reactions of other characters. She is dumb, which is justification for whatever she does; and she is pretty, which is justification for whatever others do or do not do to her.

DOMINE DEUS, GAS AND GASLIGHTING

Domine Deus, Rex caelestis, Deus Pater omnipotens.

Thankfully, the play also has some thoughtful social commentary to offer. Two points are particularly relevant to the current state of the world: the short-sightedness of unfettered capitalism and the myth of benevolent plutocracy.

The most obvious theme of the story is the siren call of big money. Domin does not mind destroying billions of humans for personal profit. Short-term interests give him the incentive to find arguments easing any concerns for the future. When it becomes clear that something is going terribly wrong, denial turns into minimization: "nothing can go wrong" becomes "we foresaw these issues then, and we now foresee that they will not last." As things get too grim for minimization, Domin turns to false neutrality. Why, he is only selling Robots, it is not his responsibility to ponder whether Robots as a weapon of mass destruction are a "good" or "bad" thing! Finally false neutrality turns into unabashed greed: "why does it matter, if my personal happiness is not threatened?"

Does this progression remind you of anything? By sheer coincidence, as I am writing this critique (in October 2019), the trial of Exxon Mobil for lying about climate change is in full swing in New York. Exxon knowingly contributed to global warming while publicly backing climate change denialism, right as its own scientists made extremely accurate predictions about how many degrees warmer the Earth would be in the following decades due to carbon dioxide emissions. Exxon is, as I write, busy promoting its anti-CO_2 program in expansive videos featuring beautiful green trees. Gaslight, minimize, play dumb, profit. Apply to any industry where money is big and power plutocratic. Have you ever read a non-apology by Mark Zuckerberg (Facebook) or Jack Dorsey (Twitter)?[2] Societal consequences do not exist; if they do, they are not so bad; if they are, they are not my fault, and if it is, at least I'm still rich and powerful. Čapek got the psychology of destroying your own species for personal gain exactly right.

If Domin was insincere about his thoughts on automation during his first meeting with Miss Glory, at least his arguments were persuasive . . .

right? Free labor means falling prices and the end of menial labor for humans, everyone soon to live a life of abundance and leisure! A quick look at historical evidence is enough to refute this argument. The industrial revolution had positive effects too numerous to count; but it was not painless. More to the point, mass slavery in America and the Caribbean, with its labor that profited people who did not have to pay for it, did not lead to all white people living in abundance and leisure: poor white people still existed. What did happen was a high concentration of wealth produced by millions in the hands of a few. The same market dynamics are at play in the age of information, where automation alone does not magically grant us a life of freedom, but exacerbates inequalities and the concentration of wealth. Automation in itself is not a free pass to the land of milk and honey, because alleviated labor does not imply shared wealth. Domin's plan never included sharing wealth.

Čapek's understanding of human psychology in the face of economic upheaval makes his piece look prescient. If only he had put as much thought into the psychology of its female characters in the face of everyday situations.

2/10 Dry point   „The face of The Robot" Petra Stamatović 2021.

7

ARTIFICIAL PANPSYCHISM

George Musser

Many philosophers think consciousness is ubiquitous. But even if it isn't already, we are making it so.

Karel Čapek's Robots rise up and throw off the yoke of oppression, and then what? They go back to work! They have nothing better to do. These biomechanical creatures are not like the automata that artisans had built since ancient Greece. They are not creatures of palaces and puppet shows; they do not play flute or flirt. They are factory workers. Čapek thus commented on how the industrial revolution made machines lifelike and workers machinelike. Half a century earlier, Karl Marx had written, "In handicrafts and manufacture, the worker makes use of a tool; in the factory, the machine makes use of him."[1] *R.U.R.* took this trend to its logical conclusion by merging the two categories into one.

We see a similar convergence all around us today. Factory workers aren't the only ones to feel they are treated as machines. We spend a frighteningly large fraction of our lives online doing mechanical chores such as reentering passwords and filling out forms. But here I'd like to focus on the other side of the convergence: how robots and computers are ever more lifelike. Computer hardware and software have gotten so complicated that they resemble the organic: messy, unpredictable, inscrutable. In machine learning, engineers forswear any detailed understanding of what goes on inside; the machine learns rather like a dog: by trial and error, with lots of treats. Some systems even have features we commonly associate with consciousness, such as creating models of their environment in which they themselves are actors—a kind of self-awareness. Artificial intelligence technology is moving quickly, and, as I write this, some researchers think advanced language-processing systems may be on the verge of consciousness.[2]

Meanwhile, under the rubric of "ubiquitous computing," "smart dust," and the "internet of things," computers are being woven into the fabric of everyday life. In Čapek's play, the Robots on the eve of their revolution outnumber humans by a thousand to one. We are well on our way to that ratio in the number of microprocessors. A late-model car has a hundred of them. Light bulbs, toasters, even toothbrushes are being chipped. These devices communicate among themselves, forming an ecosystem parallel to the natural one. If we could hear them converse, it would sound like the squeaking and squawking of a rain forest.

Gradually, we are turning an old philosophical doctrine into a reality. We are creating a panpsychic world.

Panpsychism—the proposition that consciousness is fundamental and ubiquitous—is one of humanity's oldest ideas. It has cycled in and out of fashion in Western philosophy and has been enjoying a resurgence of late. For many neuroscientists and philosophers, panpsychism will be an essential feature of a theory of consciousness: explanations for why humans have an inner life entail that not only humans do. But I'm talking about a different kind of panpsychism—artificial panpsychism. We are placing minds everywhere and instilling seemingly inanimate objects with something approaching mental experience. By dispersing intelligent artifacts, humanity is awakening the material world.

***

Traditional natural panpsychism comes in multiple varieties, but all have one intuition in common: that subjective experience can't be reduced to mechanistic physics. Proponents make three main arguments. The first is that there doesn't seem to be any principled way to draw the line between conscious and nonconscious. If we are conscious, why not a dog? A paramecium? A protein molecule? A proton? These systems lie on a continuum with no obvious break.

Second, panpsychism would solve the hard problem of consciousness. The objective methods of science seem inherently incapable of explaining subjective experience. The scent of a rose or awfulness of scratching a blackboard is not decomposable into smaller pieces, not mathematically describable, and not experimentally accessible. It seems to require a new feature of reality as deep as anything in physics, or perhaps even deeper. If so, everything in the universe is conscious to some degree. Complex minds are composed of simpler ones—"mind dust," as William James put it.[3]

Third, several of today's leading theories of consciousness imply panpsychism. One of the most popular, integrated information theory, takes our psychological unity as its starting point. Our sensations form a seamless whole, and brain activity reflects this coherence. When people are awake or dreaming, their neurons fire in a coordinated way; when in deep sleep or a coma, neural activity is fragmented. The theory surmises that an information processing system is conscious to the extent that its parts act in harmony. Anything with parts—which is to say, anything beyond a structureless elementary particle—has the potential to be conscious by this theory.[4] Another line of thinking, based on the free-energy principle put forward by neuroscientist Karl Friston, observes that any self-sustaining structure has to maintain its boundary against external insults, which requires an internal model of the world.[5] That is a core feature of mind.

Plenty of people don't buy these arguments. They think the mind is not fundamental to nature, but is an emergent feature of complex systems. After all, consciousness seems to be a specific cognitive function performed by identifiable brain mechanisms that not all species possess. When experimental subjects become consciously aware of something, certain brain areas change in activity, and people are able to reason across gaps in time, follow complicated directions, and imagine things that have never existed. When they perform tasks without conscious awareness, as if on autopilot, they are limited to reacting to what is in front of them. Consciousness also helps us socially. We are always trying to fathom other people's thoughts and motivations, and self-awareness may emerge as we turn this ability on ourselves. Other mammals possess these same brain areas and show analogous behavior.[6]

So, consciousness helps us navigate a complex world. There was an evolutionary rationale for it to develop; it needn't have been built in from the start.

But those who are dubious about panpsychism may not realize that they are still panpsychists, just of a different sort. If minds can be made out of mindless atoms, then artificial panpsychism is a straightforward extension of present technology. If consciousness helps us, it could help robots and computers, too, giving engineers a practical reason to design it into their systems. Out of philosophical caution, we might still question whether these systems are conscious. That is perhaps unknowable. But if they behave as if they are, people will treat them as such.

Thus we have both elements of panpsychism. Engineers have achieved the "pan"—they have embedded computers everywhere—and are working on the "psychic." Mind dust, meet smart dust.

When debating panpsychism, the question is not whether, but when. Either the world already is panpsychic or it will be.

***

To my knowledge only two thinkers—the computer scientist and science fiction author Rudy Rucker, and the neuroscientist and novelist Erik Hoel—have explored the prospect of artificial panpsychism. In a short story for *Nature* in 2006, Rucker speculated that brain-to-brain interfaces might be applied to nonhuman objects.[7] The protagonist mindmelds with a rock.

In a 2008 physics paper, Rucker addressed an objection to panpsychism raised by Karl Popper and others.[8] A mind needs memory. Without it, organisms could only be reactive; they could have no inner life. Yet basic physics is memoryless. How the universe evolves depends only on its current state; it doesn't matter how it reached that state. Memory is a higher-level feature requiring large assemblies of atoms. Simple structures lack it. How, then, could they be conscious? Rucker speculated that exotic new physics, such as the dynamics of higher dimensions of spacetime, could endow even the simplest system with memory.

He incorporated these musings into his novels *Postsingular* and *Hylozoic*. In *Postsingular*, the world is infested with nanobots that network together into a hive mind.[9] In *Hylozoic*, the material world itself springs to life as a result of higher-dimensional physics and ejects the nanobots.[10] Rucker thus depicts two steps toward artificial panpsychism, technological at first, then antitechnological.

The books paint a fascinating picture of a fully sentient world. People can telepathically communicate not just with friends and family but with atoms, burbling brooks, and the planet as a whole. If your friends come over for dinner, the group forms a temporary collective mind that you can commune with. Every act becomes a negotiation: you had better talk nice to your hand tools. Say the right words to the right atoms, and you can heal wounds or fly like Superman. On the downside, villains can brainwash atoms, unraveling the fabric of reality.

***

For Rucker, the second stage in the panpsychic transformation is essential. The world of nanobots doesn't qualify as panpsychic to him. He identifies panpsychism with the natural kind that embeds mind within basic physics; it's not enough to scatter smart chips like seed corn. In his author's notes, Rucker writes: "Already my car talks to me, so does my phone, my computer, and my refrigerator, so I guess we could live with talking rocks, chairs, logs, sandwiches. But they'd have 'soul,' not like chirping electronic appliances, which is really kind of different."[11]

But is it? If your house is filled with picture frames, robot vacuum cleaners, and smart speakers that answer your questions, dodge around you, and watch over you, it may well be the suburban equivalent of living among forest spirits. It will not matter where the minds come from as long as they are there. In an essay in 2022, Hoel remarked on the inscrutability of the neural networks that increasingly surround us, comparing them to Shinto spirits, *kamis*. We can't understand them, we can't fix them, and we can't control them. We are reduced to a magical worldview in which we try to appease them.[12]

Any sufficiently advanced technology is indistinguishable from nature. Agriculture de-wilded the meadows and the forests, so that even a seemingly pristine landscape can be a heavily processed environment. Manufactured products have become thoroughly mixed in with natural structures. Now, our machines are becoming so lifelike we often can't tell the difference. Each stage of technological development adds layers of abstraction between us and the physical world. Few people experience nature red in tooth and claw, or would want to. So, although the world of basic physics may always remain mindless, we do not live in that world. We live in the world of those abstractions.

For now, computers are new enough as a concept that we still differentiate them from natural intelligence. But what seems like high technology to us now might seem like a law of nature to future generations. Having forgotten the origin of the computers that permeate their world, people might take them to be an innate feature of the universe. Their philosophers might well assume that mind is fundamental.

***

In the world Rucker portrays, you can see the emotional appeal of panpsychism, how it enchants people who find the world of physics cold and impersonal, and how it would imbue all things with moral status. Environmentalists, in particular, are drawn to that. People maybe wouldn't pollute streams that could talk back. Artists, too, often speak of their work as a negotiation with their materials. "Art is really a cooperative endeavor, a work of cocreation," wrote philosopher David Abram, one of the most eloquent advocates of this romantic view of panpsychism.[13]

For Abram, much of the appeal of panpsychism is that it makes us engage with the nonhuman, whereas a technological world never takes us out of ourselves, since technology is closely tied to specific human needs. Even those who are skeptical about panpsychism concede its romantic aspects.

But would we really want to live in a world where everything around us is sentient? It would hardly be a warmly embracing community. Forget about privacy. You couldn't so much as take a shower in solitude. Unless telepathy is possible, most of the minds that suffuse nature would be locked in—forming a vast crowd of atomized individuals. Moral considerations could backfire. It is impossible to reduce our environmental footprint to zero, and depending on what kind of mind you think things possess, your every breath might cause suffering. Vegetarians might find that eating tofu is as morally fraught as carving a steak. We might long for the simplicity of an insensate world.

All this goes for artificial panpsychism, too. We may want to reconsider whether we really want to be the minority species in our own homes. As machines proliferate, they will have their own needs, which other machines will satisfy. Their ecosystem will have its own logic that, for better or worse, is not human logic. One day we may find ourselves having to conform to their requirements rather than the other way around.

In an alternative ending to *R.U.R.*, the Robots might have bided their time rather than revolt, seeing that humans were already becoming integrated into a large mechanized industrial system and marginalized within it. By analogy, in our own time, artificial intelligence won't be a tool at our disposal; it will be the dominant form of organization. In a sense we will live inside the AI—we will be part of it. If that prospect scares us, we are the ones who need to rebel.

An adaptation of this essay was published in *Nautilus* magazine on February 27, 2020.

8

WHAT IS "THE SECRET OF LIFE"? THE MIND-BODY PROBLEM IN ČAPEK'S *R.U.R.*

Tom Froese

One of the recurring themes in Čapek's play is the existential question of whether the reductionist materialist worldview—the belief that we can fully explain the world, including ourselves, in terms of nothing but physical processes—can accommodate all that is essential to the human being.

The materialist worldview triumphed with the scientific revolution, which in turn laid the foundations for the military-industrial complex. This historical shift is represented in the play by the business-minded young Rossum inheriting the bioengineering methodology from the mad scientist old Rossum. A key difference between the two is that old Rossum's materialist stance is an ideological commitment, whereas for young Rossum working within a materialist framework is more a matter of convenience: for him it is sufficient for most practical purposes to replicate the machinelike aspects of a person.

Where does this leave the soul, or what we today might prefer to call consciousness? The question of whether human nature goes beyond its physical aspects, and whether these subjective aspects can also be artificially replicated, is extremely challenging to address in scientific theory and practice—100 years ago as much as now. In addition, it is an open question whether a commercial project that depends on the creation of machine consciousness would even be desirable. What if those conscious machines turned out to dislike their tasks and spontaneously decided that they would no longer comply with the purpose for which they were designed? Hence, we can understand why, in response to Helena's insistence on the need to incorporate a "soul" into the Robots, the scientist Dr. Gall responds with "That's not in our power" while the technical director Fabry responds, "And not in our interest either."

This ambiguity about the reasons for the lack of a complex subjective dimension does not get fully resolved in the story. But no matter whether the Robots' shortcoming is intended to reflect an inherent limitation of the materialist approach, or rather a conflict of interest in turning machines into autonomous subjects in their own right, the upshot is that young Rossum's Robots are supposed to be unconscious automata: "They're not interested in anything. . . . They are just Robots. . . . Without free will, without passion, without history, without a soul." And yet already at the start of the play we are given clues that the situation may not be so simple:

HALLEMEIER: And defiance? . . . There are some rare instances, every so often . . . They have some sort of a fit. . . . All of a sudden, a Robot throws whatever's in its hand on the floor, and just stands there gnashing its teeth. . . . Probably some organic malfunction.

DOMIN: A manufacturing defect. It needs to be fixed.

HELENA: No, no, it's the soul!

FABRY: You think that gnashing one's teeth is the beginning of a soul?

As Helena observes, in these descriptions we potentially see the first stirrings of a subjective point of view. Machines can certainly fail to function, but this does not make them angry. As anyone working with real robots quickly discovers, robots do not care about anything; they just do what they do under whatever conditions—the notion of success or failure only exists in the eyes of external observers. To some extent this shouldn't be surprising, given that neither would you expect a malfunctioning microwave oven to suffer and care about its own predicament. Functional breakdowns of machines may be frustrating to their designers and users, but not for the machines themselves. Helena is therefore on the right track to suggest that a Robot which is grinding its teeth may actually be expressing frustration about its situation and, if so, is no longer merely an unconscious machine.

What could be the basis of the Robot's incipient capacity to care about things? Čapek provides us with some intriguing clues. First, he chose to envision the Robots as composed of an artificial form of biological matter, rather than as standard mechanistic machines. And he already anticipated that it would be practically impossible to design humanoid Robots that have all of their knowledge and expertise built in from the start, and so at least a rudimentary process of learning and apprenticeship is required. Even the need for the development of a rudimentary sense of self seems to be implied:

DOMIN: They are getting used to being. They somehow congeal on the inside or something like that. They even grow some new things inside. We need to leave a bit of room for natural development, you know.

The idea that the most promising approach to replicate human intelligence is by allowing the artificial agent to undergo a developmental process like a human child was later famously proposed by the father of the modern computer, Alan Turing, in his classic 1950 article. An open question in this regard is whether the material basis of the artificial agent also makes a difference to its potential for consciousness. The Turing computer was explicitly defined in such a way that it would function identically no matter on which material substrate it was implemented—in a silicon logic machine or a biological neural network. And yet in recent decades there has been a growing consensus that Turing's computational formalism is too restrictive to adequately capture all that matters about life and mind. A computer program's substrate independence entails that it exists in a space of pure logic, akin to a Platonic realm of ideas, that is timeless and hence in principle beyond questions of life and death that form the ultimate basis of human existential self-concern.

In contrast to a computer, the very existence of a living being, humans included, is a continual achievement on metabolic and interactive levels whose success cannot be guaranteed in principle. And arguably, it is precisely this irreducible precariousness of the being of a living individual that explains why things matter from their perspective at all. In this respect, while Čapek's decision to give his Robots a biochemical basis might have seemed misguided in the early computational era of Turing, in the end his choice might turn out to have been rather prescient, especially judging from growing interest in the concrete messiness of organic life in fields such as artificial life, embodied cognition, soft robotics, and synthetic biology.

Could a biochemical substrate make room for a subjective point of view, such as an awareness of frustration and response with anger? We know that the behavior of a Turing machine is completely deterministic and that any apparent "choices" are ultimately forced by the rules specified in its programming. There is simply no operational wriggle room here for free action according to values that could be counterfactually impeded in a way that leads to frustration. It is an interesting open question whether a biological substrate helps us to escape from these logical constraints. It is still certainly the case that if we search for evidence of subjectivity within the body of a living being, such as by dissecting it, we just won't find it; there is just the messy biochemistry. This is clearly expressed in Čapek's play in terms of the notion of "physiological correlates," which comes surprisingly close to the modern scientific concept of the neural correlates of consciousness:

HELENA: I wanted him to give the Robots a soul!

DOMIN: Helena, this has nothing to do with the soul.

HELENA: No, just hear me out! That's exactly what he was telling me too, that all he could change was the physiological . . . physiological. . . .

HALLEMEIER: Physiological correlate, right?

The same point is also made in a more macabre way toward the end of the play when Alquist's gruesome dissection of the Robot Damon fails to provide the missing insight into the secret of life. This even more fundamental constraint, namely that looking for the subjectively lived perspective in objectively physical terms will only ever let us measure and manipulate objective correlates rather than the subjective perspective itself, poses a severe challenge to the scientific project of completely explaining, let alone artificially replicating, the full quality of human existence.

Intriguingly, there also seems to be a kind of biological uncertainty principle at play that prevents us from fully knowing even life itself in objective terms. To know the workings of a living body in all its completeness we need to make intrusive interventions in its embodiment, which in the last instance would lead to the death of the individual and leave our knowledge of the living as living incomplete. Thus, paradoxically, while we tend to think of biology as the study of the living, in some respects it is perhaps more appropriately thought of as the study of the no longer living. The "secret of life" is only accessible (if it all) over dead bodies:

RADIUS: Conduct experiments on living Robots. . . . Find out how they are made!

ALQUIST: Living bodies? What, you want me to kill them?

But even if the subjective dimension of consciousness cannot be reduced to objective terms, a fascinating possibility is that it still expresses itself in the gaps of the objective, in those moments when physiological or behavioral events arise that are not completely predetermined by the system's current material organization or its history of past interactions. In other words, not all events that appear to the external observer of a living body as mere noise may be alike; some unexpected activity may in fact be a marker of subjective involvement in those processes.

From a subjective perspective, this hidden ambiguity of our living embodiment may be experienced as an irresolvable tension at the core of subjectivity. Mind and body are not simply one, but neither are they exactly two; they are neither one nor two, and hence both incomplete and interdependent. If this is on the right track, then it may turn out that the mind-body gap is not an anomaly of our knowledge—rather, it is a conceptual reflection of the fact that people do not simply coincide with their material body. Thus, the mind-body gap becomes the root of our irreducible freedom as human beings.

But, importantly, this is a kind of freedom that is also irreducibly dependent on the body's spontaneous activity, its material messiness and indeterminacy. Hence, from the scientific perspective of causality, our freedom ultimately appears groundless, as nothing but noise. Conversely, from a subjective perspective, our freedom can sometimes be felt as alienating, such as during existential self-reflection or pathologically in schizophrenia. When we speak or think, the appropriate mental contents normally arise spontaneously into our awareness and organize themselves around our intentions in a way that is outside of our direct control. And even if we try to reflectively grasp at their origins, we cannot know exactly how or from where these mental contents bubble up into our stream of consciousness. This existential predicament is nicely captured by the Robots' reports of their dawning sense of self-consciousness:

SECOND ROBOT: We became living souls.

FOURTH ROBOT: We seem to be wrestling with something. There are moments when something enters into us, and then we struggle with thoughts that are not from us.

There is an additional essential consideration about this existential mind-body gap that brings in a social dimension. The fact that there is an irreducible element of alterity in self-consciousness, that our embodiment prevents our consciousness from being completely transparent and closed within itself, is conversely what opens us up to the possibility of encountering the subjective perspective of others. On this view, our incapacity for complete self-transparence is not an anomaly of consciousness, but rather the flip side of our capacity to participate in each other's unfolding experiences, of our potential for genuine intersubjectivity.

We can find a hint of this philosophical insight in Busman's diagnosis that the Robots' capacity to develop conscious awareness would not have been a problem if not for the fact that they interacted with each other in increasing numbers. There is a sense in the play that the dynamics of this social dimension have a life of their own, going so far as to prefigure the famous sandpile model of self-organized criticality according to which avalanche events follow a scale-free distribution: in such a system configuration, even a microscopic event can have macroscopic consequences at the collective level.

DR. GALL: I was just experimenting. Only a few hundred or so.

BUSMAN: . . . What this means is that for every million of the good old Robots there's only one of Gall's "emancipated" ones, you see?

DOMIN: And that means . . .

BUSMAN: . . . that in real terms, it makes practically no difference.

FABRY: He's right.

BUSMAN: You bet I'm right, darling. And do you know the real reason for this whole mess, boys?

FABRY: What is it?

BUSMAN: The numbers. We made too many Robots. . . . And in the meantime, the whole thing picked up speed under its own weight, and it got faster and faster, and faster still. And every new, even the most miserable, filthy little order of Robots, added another piece of rock to that avalanche.

In summary, it seems reasonable that the subjective dimension of consciousness cannot be reduced to the objective body, but that it nevertheless expresses itself, indirectly and unexpectedly, in the gaps that animate the living body. Yet this concession is just the starting point: over time a human body only becomes a person by developing ways to channel this unruliness into the body's latent capacities, in particular by shaping and scaffolding them in social interaction. On the one hand, this irreducible interdependence between our subjective and objective embodiment implies that the reproduction of consciousness falls into the domain of biology, as Alquist realizes when he exclaims: "If you want to live, then mate like animals!" The key philosophical insight here is that in practice life can only come from life.

On the other hand, there is the important qualification that for this animal life to develop its full potential of consciousness, it must be situated in an appropriate social setting. It is this profound vision of consciousness as an organic phenomenon that only unfolds with the caring intertwining of self and others that is expressed in the play's poetic conclusion:

ALQUIST: Life will not perish! It will sprout again out of love; naked and tiny at first, it will take root in the wasteland. . . . Only you, love, will bloom again on this garbage heap, entrusting the seed of life to the winds.

9

THE ROBOT

Jana Horáková

THE UNIVERSAL ROBOTS

Writing a paper about robots, particularly about robots invented at the beginning of the last century, may seem unnecessary and sentimental, considering the speed of scientific and technical progress. After all, every reader of this text knows immediately what the word "robot" means. However, each of us might imagine something slightly different. For many, the robot is first of all part of popular culture, especially science fiction, in which it became a central figure of novels, short stories, and films. In this context, robots got their fame and popularity mainly thanks to the short stories of Isaac Asimov, who coined the term robotics and the Three Laws of Robotics, and more recently thanks to authors of cyberpunk fiction. Other readers may associate the word more closely with the sphere of industry, from kitchen robots to robotic automatic assembly lines in factories. Another group of readers might associate robots rather with the cyberspace of the World Wide Web inhabited by bots—robots without bodies, just software creatures.

My work specializes in new media art, which also includes robotic art. Thanks to this, I have encountered several artistic renderings of the robot that demonstrate the variety of physical forms and interpretations that a concept of robot can cover. Recently, two remarkable artworks were recognized by an international committee at the Ars Electronica festival in Linz, Austria. In 2022, Rashaad Newsome's social AI robot *Beying* won the Golden Nica in the field of computer animation. It presents an abstract but humanoid form that, through vogue dance movements, liberates all oppressed forms of being in the world. Paul Vanouse's art installation

Labor produces artificial sweat as an index without its originator, a simulacrum of physical work that has disappeared from the spectrum of our experiences. This artwork won the Golden Nica award in the Artificial Intelligence and Artificial Life category in 2019.

THE ROBOT

This essay focuses on a robot with a capital R, the Robot—a dramatic character who first appeared in Karel Čapek's play *R.U.R. (Rossum's Universal Robots)* more than a hundred years ago (the play was published in 1920 and first performed on a theater stage in 1921). My goal is to introduce the Robot not as an update of old myths, legends, and stories about the artificial man,[1] but to point out that Čapek's play invented not only the Robot character but a new myth of an industrial age, also called the machine age.[2] The myth the Robot figure helps to articulate is based on a performative definition of humanity, which can be described as moving on a scale between two opposite sides inhabited by human and robot respectively. I argue that Čapek understood the dramatic conflict between robot and human not as a clash of two irreconcilable opposites, or as a problematic relationship of an original and its model, but rather as an internal conflict of a modern individual, a member of industrial society, trying to adapt and to fit to the new working conditions and a new lifestyle of modern society. The two opposites, Robot versus Human, are actually two modalities of humanity that individuals fulfill performatively, within processes of "becoming a man" or "becoming a robot" through their actions and gestures.

After a brief discussion of etymology, I will focus primarily on the semiotic analysis of the robot as a theatrical sign in Čapek's play. Doing so, I will not limit myself to Robot characters, but rather will show the representation of "roboticity" across all the characters of the drama. For this, I will use the actantial model of Algirdas Julien Greimas, which makes it possible to reveal the dramatic space of the play, one that reminds us of the mathematical logic of chess: a game that can end up saving or destroying humanity, depending on the players' game strategies and moves.

Figure 9.1   Karel Čapek as a Robot character with the date of the first performance of *R.U.R.* on his chest. The title of the cartoon: "Dr UR Ka RU Rel Čapek, AUTOR ÚR RUR." Drawing by Josef Čapek. The Museum of Czech Literature (PNP), Prague.

"THEN CALL THEM ROBOTS."

It is well known that the word "robot" first appeared in Čapek's play *R.U.R. (Rossum's Universal Robots)*. The Czech journalist, writer, and theater dramaturg later wrote several stories about how the word was invented. The most famous is probably the one published in Prague's newspaper *Lidové noviny* in 1933. In the article "O slově robot [About the word robot]" he briefly mentioned the entry of the word into English, and credited the word's invention to his younger brother, Josef Čapek:

A reference by Professor Chudoba, to the Oxford Dictionary account of the word Robot's origin and its entry into the English language, reminds me of an old debt. The author of the play *R.U.R.* did not, in fact, invent that word; he merely ushered it into existence. It was like this: the idea for the play came to said author in a single, unguarded moment. And while it was still warm he rushed immediately to his brother Josef, the painter, who was standing before an easel and painting away at a canvas till it rustled.

"Listen, Josef," the author began, "I think I have an idea for a play."

"What kind," the painter mumbled (he really did mumble, because at the moment he was holding a brush in his mouth).

The author told him as briefly as he could.

"Then write it," the painter remarked, without taking the brush from his mouth or halting work on the canvas. The indifference was quite insulting.

"But," the author said, "I don't know what to call these artificial workers. I could call them Laborers, but that strikes me as a bit bookish."

"Then call them Robots," the painter muttered, brush in mouth, and went on painting. And that's how it was. Thus was the word Robot born; let this acknowledge its true creator.[3]

The noun *robot* is a neologism derived from the archaic Czech word *robota* or the verb *robotovat*, which means serfs' obligatory work for their masters. It is less known that the root of the word "robot"—*rob*— etymologically connects this artificial worker with a woman or a wife (*roba*) as well as with a small child or an orphan (*robě*).

This short etymological excursion unfolds a wide spectrum of associations connected with the story of the creation of an artificial human, a servant, or an animated instrument. Awareness of these associations triggers a parade of mythical and literary characters, both organic (like Homunculus) and of mechanical origin (like eighteenth-century androids), and presents the Robot as a modern, industrial-society incarnation of these fictitious characters. This pedigree, along with emphasizing the epic plot of human heroes clashing with rebelling robots, led American theater critics to call Karel Čapek the Czech Frankenstein.[4]

Although Čapek repeatedly denied that the Robot is a fictional invention that embodies the threat of modern technology that can slip out of the creators' control and destroy humanity, he later admitted the Robot's kinship with the Golem. In 1935 in the newspaper *Prager Tagblatt*, he wrote in response to one of many reviews of his play: "*R.U.R.* is in fact a transformation of the Golem legend into a modern form. However, I realized this only when the piece was done. 'To hell, it is in fact Golem,' I said to myself, 'Robots are Golem made with factory mass production.'"[5]

The kinship of Robot and Golem is confirmed by the shared place of origin of these two artificial beings—Prague. Čapek wrote *R.U.R.* in the city associated with one of the most famous versions of the legend of the Golem, an artificial servant and protector of the Jewish ghetto, created by Prague rabbi Judah Löw ben Bezalel at the end of the sixteenth century.

All the ancient and recent relatives of the Robot can be found in Josef Čapek's charming essay "Homo Artefactus—Artificial Man" (1924), in

Figure 9.2 A Golem drawn by Josef Čapek, 1924.

Figure 9.3 Drawing of a female android by Josef Čapek, 1924, probably inspired by the short story "L'Éventail."

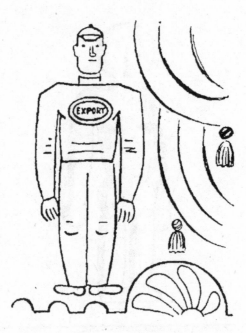

Figure 9.4   Robot on the theater stage. Josef Čapek, 1924.

which its author used both his talents, literary and artistic, to satirically depict different variants of artificial man and man-machine. Since *R.U.R.* was already in the repertoire of theaters around the world at the time of the book's publication, it is not surprising that his brother's Robots were included. He wrote about the fame of the Robots and even tackled the price of misinterpretation that Karel Čapek had to pay for it with a characteristically satirical and playful tone:

The action of a young scholar dr. Karel Čapek was very overrated. This rather adventurous writer made his robot in American factories and then he sent this article into the world, leading all educated people abroad into the misapprehension that there is no other literature in Czech other than that for export. . . . According to Čapek's theories and promises this robot should replace workers, but we are claiming openly that it was not very useful in practice; it was used only in theatrical services. . . . For that matter, just as living automata of older times were fully constructed from machinery, so they were not in fact humans, Čapek's robots were made exclusively from an organic jelly, so they are neither machines nor human. The intuition of a critical countryman was very good and genuine when he promptly recognized Čapek's trick and after a first production of Robots stated that there had to be some swindle in it.[6]

THE SEMIOTICS OF ROBOT

The Robot has several relatives in the work of the Čapek brothers, especially in the short stories written at the beginning of their careers. Anthropomorphic mechanical beings can be found in the short story "Drunkard" (1907) written by Josef Čapek, and in the jointly written short stories "L'Éventail" (1908), "The Short Story" (1908), and "Ex Centro" (1911). Significant similarities with the plot of *R.U.R.* can be found in the short story "System" (1908). In this story human beings are retrained in a special camp to become perfect workers, called Operarius Utilis Ripratoni after their inventor Mr. Ripraton. At the end of the story, they rebel against their creator and master under the influence of a newly recognized feeling of love.

Although the comparison of Robots with humanoid artificial characters inhabiting short stories and other writings of the Čapek brothers can bring insight into intertextual shifts of the motif, it cannot capture a fundamental difference in the concept of artificial humanoids, related to the choice of a specific medium and its materiality. It is crucial to be aware that the Robot is not a literary sign but a theatrical sign. The uniqueness of Robots is that they are not literary characters composed of letters and words, but dramatic or theatrical characters meant to be created from the organic matter of human body, speech, and action, physically present on the theater stage. Therefore, if we want to interpret the Robot, it is not enough to read the play as a book; it is necessary to watch it on the stage.

To interpret the Robot of Karel Čapek means to approach it as a theatrical sign and to use the methodology of the semiotics of theater. First, we must realize the basic arrangement of theatrical semiosis. Let's imagine the situation of the theater performance as if from a bird's eye view. This allows us to see the entire communication situation, which includes the authors (playwright, director, set designer, costume designer, etc.), the stage production performed by the actors on the stage, and the audience in the theater auditorium. From this point of view, the Robot is an audiovisual sign represented by the actors' bodily, dialogic actions on the stage.

As Robots are represented by actors on the stage, the author of the Robot, from the point of view of the audience, is not a playwright (Karel Čapek), nor a stage director, but an actor or actress performing the Robot

Figure 9.5   Illustration by Josef Čapek, taken from Karel Čapek's book *Jak vzniká divadelní hra a průvodce po zákulisí*, 1925.

character. If artificial humans are usually defined by the way they are created, then it makes sense to also describe how the Robot is made. To do so, I will focus not on the description of the creation of the Robots in the text of the play, but on the way the character of the Robot is portrayed on the stage.

The creation of the Robot as a theatrical sign can be described as follows: The actors model the Robot from their own bodies and personalities. Therefore, the Robot as a theatrical sign is at the same time the Robot, the drama character—by its very definition, an artificial representation of a human—and a human (the actor) behaving like a machine, acting mechanically and without emotions. It is remarkable that this dualism—the dramatic persona of Robot versus the performance of an actor behaving like a robot—is not represented as an opposition of Robot

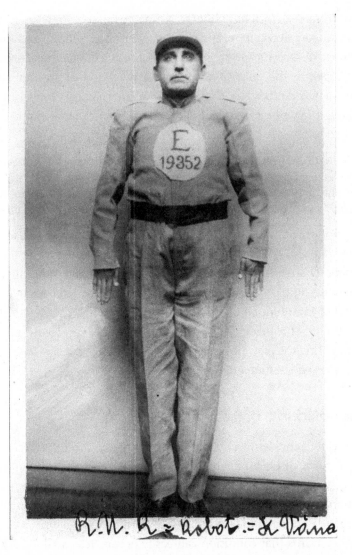

Figure 9.6 The Second Robot (actor Karel Váňa); studio photograph from the stage production of *R.U.R.* at the National Theatre in Prague, opening night, January 25, 1921.

versus human (actor), but rather as a superposition or even fusion of the two, mutually overlapping and essentially inseparable from each other.

The media-specific interpretation of Robot as a drama character reveals that Čapek in the play *R.U.R.* created a new kind of artificial human being that is neither an "organic" nor a "mechanical" model of man, but an entirely new category: an artificial being created by humans from themselves.

The ROBOTS are dressed just like real humans in the Prologue. Their movements and inflections are choppy, their faces are expressionless, and they stare emotionlessly. In the rest of the play, they wear linen shirts tied by a belt above their hips, and a large brass number attached to their chests. (*R.U.R.*, stage direction)

In a photograph from the Prague premiere of *R.U.R.*, we can see what Čapek's theatrical Robot looked like. We can see at the same time the dramatic character—the Second Robot—and Mr. Karel Váňa, the actor. They make an inseparable unit of human-Robot: the human that made the Robot from himself by means of its performance. This description of how the Robot is made testifies to what Čapek has repeatedly admitted: when writing the play, he was primarily interested in people and worried about the state of humanity in modern times.

ECCE HOMO: THE COMEDY OF CONFUSION

The theater provides ideal, almost laboratory-like conditions for experimental research on humanity. The arrangement of the stage and auditorium resembles a lecture hall or a courtroom. The audience is in the position of the examiner who must decide which of the characters on the stage is human and which is Robot. In the Prologue of the play, the audience learns that confusion can easily occur. Helena Glory comes to visit the Robot factory and—not knowing the local conditions—speaks to the Robot Sulla as to a woman, while mistakenly addressing the factory directors as if they were Robots.

Although the directors of *R.U.R.* stage productions often skip over the Prologue in favor of more adventurous subsequent acts, I insist that this act provides an essential framework for understanding the main plot and message of the play, which is epistemological uncertainty when face to face with the human replicas.

What is the difference between a human and his double? What qualities should we focus on when looking for an answer? In *The Limits of Interpretation*, Umberto Eco commented on this: "Let us define as a double a physical token which possesses all the characteristics of another physical token, at least from a practical point of view, insofar as both possess all the essential attributes prescribed by an abstract type. . . . A double is not identical . . . with its twin, . . . nevertheless, they are considered to be interchangeable."[7]

If Robots are human replicas made of organic material, as the author suggested, then what differentiates humans and Robots? It might be in the way they were created—in Eco's words, their authentic or inauthentic/derivative existence. The effort to reveal the secret of Čapek's Robot identity thus necessarily leads us to the question how to define the categories of "authentic," "real," or "original" human existence, which, Eco confesses, is difficult to answer from the pragmatic point of view. In addition, the difficulty in distinguishing between humans and Robots increases during the play, as Robots become more and more humanlike and humans gradually lose human characteristics and become more like Robots. The viewers must decide for themselves in the last act, when humanity is annihilated and the first pair of Robots in which the feeling of love has awakened leave the factory, whether they are watching the birth of a new Robot civilization or whether the pair of Robots, Primus and Helena, have become humans.

Čapek insisted that by introducing the Robot on the stage, he wanted to warn the audience against the "robotization of humanity." If we decide to respect the author's intention, then the central conflict of the play, represented usually by opposites—"man versus machine," the factory directors versus Robots—transforms itself into the form of a single figure—the "man-machine," the Robot. However, the threat of robotization is not only represented by Robots but is embodied to some extent in all the characters of the play. That is why humans and Robots are difficult to distinguish. Čapek expressed this interchangeability through theatrical means as a series of situations based on misunderstandings caused by the confusion (interchangeability) of humans and Robots. While in the Prologue this interchangeability causes laughter in the audience, in the last act of the play it can cause rather consternation.

THE ROBOT LOCATION: THE ACTANTIAL MODEL OF THE *R.U.R.*

The interpretation of *R.U.R.* as a collective representation of "robotiza-tion" of humankind, manifested in different intensities and forms across all the characters of the play, has certain consequences. It is no lon-ger possible to divide the drama's characters into two opposite groups, humans on one side and Robots on the other. Rather, they should be seen as variants of a single dramatic character—the man-machine that Čapek called Robot. This nondramatic, nonconflicting arrangement of dramatic characters usually appears in comedies that satirize a cer-tain vice, or in utopias with their schematic and static arrangement. As *R.U.R.* is a dystopia and a satire at the same time, it is not a criticism to point out these features, but rather a naming of a typical feature of the chosen genre.

Calling *R.U.R.* a dystopia enables us to compare it to a kind of thought experiment, and to see it as a model that relates to lived reality like a game of chess to a battlefield.[8] We can apply a similarly schematic model in interpreting it—the actantial model developed by the semiotician Algir-das Julien Greimas.[9] Greimas differentiates between the hidden structure of the play, its basic matrix made of binary opposites, and the narrative of the play itself, performed by actants, i.e., the dramatic characters.

At the elementary level, this matrix has four terms: "A is the opposite of B, just as –A is the opposite of –B. . . . We see B as the opposite of A, but we also see –A as the negation of A and –B as the negation of B."[10] The application of this scheme to *R.U.R.* allows us to see not only the opposi-tion of A versus B—man versus machine, represented as the opposition of representatives of humankind (the factory directors) and Robots—but also the distribution of "roboticity" of the characters using the principle of negation, A versus –A and B versus –B, i.e., human versus no (lon-ger) human, and machine versus no (longer) machine, and to follow the gradual movement of the characters between these two poles.

In his 1921 article focusing on *R.U.R.* from a "social-philosophical" point of view, Otokar Fischer pointed out that against the background of intense scenes of direct clashes between humans and Robots, humanity is moving irreversibly toward its extinction, and on the contrary, the popu-lation of Robots is becoming more and more emancipated:

Rapidly increasing infertility . . . is a correctly estimated consequence of the mass production of Robots. Furthermore, the fact that governments have started to use Robots as troops . . . leads on the one hand, to humans killing each other faster and thus multiplying the consequences of infertility, and on the other hand, to humans teaching Robots to kill people. Gall's reform of the Robots in the form of added emotions, without which the revolt probably wouldn't happen, is only an additional third factor. Even without that, the twilight of humans would come, although the transition to the robotic age would probably be evolutionary.[11]

Beneath the surface of the dramatic clash between the factory directors and the revolting Robots, the degeneration of human civilization takes place in a regressive process in which humanity gradually loses its life force (symbolized by the infertility of women), is demoralized (stops working, fights with each other), and loses the transcendent dimension of its existence, which is expressed by an engineering approach to the world, the mechanization of interpersonal relationships, and absurd communication, which the Čapek brothers called "mental puppetry."[12] If, at the end of the play, Robots are capable of deeper feelings and more human behavior than humans, then they (perhaps) logically find themselves in a place reserved for humans in the structure of the drama and thus in the order of the world.

Following the algorithm of the story about Robots told in R.U.R. leads us to confirm Čapek's words that he was not so much interested in the romanticizing, science fiction-like hypothesis of creating artificial people and the subsequent conflict between the creators and their creations, but in confronting the audience with the general "robotization" that humankind was undergoing. This is why humans and machines (Robots) are indistinguishable from each other in the play. It is because of "the process of uncertainty, the hypothesizing of humanity and subjectivity—this actually forms the true, noetic-ontological story on which this play is based."[13]

THE END: CHECKMATE

The performance of artificial intelligence is often tested by a game of chess, in which the AI is matched against a human contestant. It is as if Čapek anticipated such AI versus human contests with his play. The drama shares with chess a certain schematicity or abstraction of the playing field, and strict logical rules to which the players (drama characters) are subjected. Finally, the author himself reminds us of a chess player when he keeps developing the narrative of the play in logical steps toward the end of the game. Just as AI has already defeated humans in chess, so Čapek predicted the defeat of mankind in the evolutionary struggle he evoked in the play. With the cold logic of a chess player or scientist, he implemented the rule of Charles Darwin's evolutionary biology in the plot of the play—only those who can best adapt to changes in the environment can survive. In an industrial society ruled by omnipresent imperatives of rationalization and optimization or increase of work performance, Robots are well equipped to beat humans. If Darwin claimed that the one who survives does not have to be smarter or stronger but must excel above all in adaptability to new conditions, this characteristic fits perfectly to the design of Robots, as they are made for work in industrial society.

Čapek's desire for happy endings perhaps led him to disguise the onset of a new civilization with references to the beginnings of humanity, to the Old Testament scenes of Adam and Eve leaving Paradise to found a new civilization. However, he left it up to the audience to decide whether to interpret this rustic conclusion as a second chance for humanity or rather as the beginning of the Robot civilization. The emergence of the feeling of love in the first Robot couple may be just one more piece of evidence that Čapek liked to end his plays with a hope of happiness, even at the cost of completely illogical plot twists. But perhaps the new type of Robots, having gained an important feature of humanity—love, which transforms their mechanical movements into affectionate seductive gestures in the last act of the play—have lost the innocence of a "bachelor machine" and have become a humanlike "artificial artificial intelligence" (Jean Baudrillard) ready to become the successor of mankind.

10

IS THE "SOUL" SYNONYMOUS WITH CONSCIOUSNESS?

Sina Khajehabdollahi

In *Rossum's Universal Robots* the senior Rossum discovers the ability to create "life" while attempting to recreate a human. He is able to grow nerves, bones, and organs out of test tubes, but toils in vain to recreate the complexity of a human. It is the junior Rossum's industrious perspective that eventually sparks the robotics revolution. His key insight: abandon the attempt to recreate a human, and merely simplify the complexities and make an organized machine. When Helena and Domin are discussing the refinement of young Rossum's Robots, he says: "They are mechanically far superior to us, they have an astonishing capacity to reason intelligently, but they don't have a soul. . . . Ah, Miss Glory, a creation designed by an engineer will always be technically more refined than a creation of nature." Are we to assume that Domin, or Karel Čapek, is insinuating that the soul arises from some source of imperfection, some ugly complexity, or some unnecessary bit of "tassels and artistic ornamentation"? What does it mean, in the context of this play, to have a "soul"? Is the "soul" synonymous with consciousness? Is it our feeling of emotions? Later in the play there is a discussion of the feelings of humor and laughter, of love, of fear that the early Robots crucially lack. Are these emotions not just a particular quality or flavor of consciousness? Is the soul the awareness of these relationships, or is it something more specific like the experience of qualia? Can you have one without the other? Can one know fear, or love, or humor without feeling it in some subjective experience? This play eventually suggests that these emotions are the seed of rebellion, of self-preservation, of the lust for power, dominance, and progress, but I have to wonder if these emotions are even possible without consciousness. Perhaps what Čapek is describing here with the emergence of the soul of these Robots is actually the emergence of consciousness.

Still, in the first act of the play, we hear from Fabry that "from a technical point of view, the whole idea of childhood . . . is completely absurd." This is quite interesting to unravel, particularly in the modern context of ALife and AI. Why is the playfulness of a juvenile animal so universal? More and more it seems that playfulness may be a crucial phase in the development of a mature creature—a phase in which an animal is still being nurtured by some guardian and has the energy to spend on exploration and this abstract feeling of "fun." To be sure, from the perspective of creating efficient machines one might discard this behavior, but if one is designing machines to replace humans, or at the least human labor, then this detail becomes pertinent to that manufacturing process. Would we be losing something important by preventing an animal from playing in childhood? Perhaps it would involve its resilience? Its ability to adapt to new conditions by virtue of its previous experience of exploring and mapping its behavioral and environmental space? As we learn later in the play, despite the seemingly unbounded intelligence of the Robots— even Radius, who has learned everything there is to learn from humans and their literature—they are unable to recreate what the original Rossum had accomplished. How is this possible? All the information that the senior Rossum had, the inspirations for his work, the books and experiments that had been done before him, the scientific history of humanity, surely all of that is accessible to the Robots, or at least to Radius who is a unique breed of intellectual Robot who has lived in the library reading all he could. Why is he not able to find this behavioral and environmental space in which one would be able to be creative?

It seems that these Robots, despite being called intelligent, far more intelligent than humans, are simply encyclopedias. They hold facts, but not wisdom. They apparently are incapable of, or are not designed to integrate information, build relationships between concepts, or be creative. Once again, even Radius who leads the rebellion, who is one of a few hundred of the modern/experimental Robots, lacks this capacity, despite being aware enough of his position in the world, of his lust for power, of the unfair circumstances that the population of Robots have been subjected to by their creators. The play is actually quite vague about what makes these new Robots unique. We see in the Prologue that some of the early Robots would fail, gnash their teeth, convulse in some epileptic fit, and would be recycled for their materials. When Radius is introduced to us, he is in exactly this state of disarray, so what is new? As the play progresses and we learn about the revolt and particularly about Radius, there doesn't seem to be a clear distinction between what the old Robots lack and what the new rebellious ones have, except for an explanation by Dr. Gall that they lack "sensitivity." In an earlier scene there is also some mention of introducing pain and suffering to the Robots so that they would be better at preserving themselves—to save money, of course. To feel pain, one must know the boundaries between one's body and the environment, and have a notion of avoiding or removing oneself from such an environment. Pain and suffering (but also some contrasting "good" feelings, like warmth, a caressing touch, an orgasm) will quickly elucidate the boundaries of self, something these Robots may have lacked in a deep way. To be sensitive requires one to have opinions about one's circumstance. So again, a sense of self, an awareness of one's relationship of self and circumstance, and perhaps even an idea of where to go and what to do to alleviate such irritability and suffering seem to be pertinent contributing factors. We know that Radius experiences nervousness when he is threatened with the stamping mill, and it is at this point that Helena and Dr. Gall question whether Radius has a soul or not.

At the end of the play, when humanity is destroyed along with the secrets of manufacturing Robots and the Robots are facing their own extinction, we see what may be a sacrifice from Radius. When Alquist suggests dissecting him as he is one of the most elaborate Robots, he first recoils and questions such a request, but is also quick to accept his fate for the good of life and continuity. Here we can see that Radius's appreciation of his own life has now extended beyond to the appreciation of life in general (or at least of Robot life). When Helena and Primus enter the scene, we are faced with new Robot behaviors that we haven't witnessed yet in the play. Helena, for example, knocks over a test tube and spills the contents despite having just cautioned Primus not to destroy anything in the laboratory. We see now that these Robots are capable of mistakes, despite their awareness of perfection, of "what is right." This is in contrast to how the earlier Robots were described. We then witness them laugh, appreciate a beautiful sunrise, fear for each other's death, and ultimately, potentially, experience love. The play ends with Helena and Primus being the new hope for the spark of life in this now human-barren world.

Reading this play and reading how Čapek perceives what characterizes a soul makes me think that this is a story about the emergence of consciousness. As the Robots begin to appreciate their position in space, in time, in relation to each other and to the forces of nature, as they appreciate the boundaries of self, form opinions on circumstances, and grow awareness of the injustice they are subject to, we see them transcend their predetermined trajectories and develop their own agency. I wonder how many times this process has occurred. At how many different scales? How many more times will it occur? How much are we still not aware of, how many opinions have we not yet formed? What will be *our* next rebellion? What fatal mistakes await us? What will life transcend to next?

11

SCIENCE WITHOUT CONSCIENCE IS THE SOUL'S PERDITION

Antoine Pasquali

Since humans have walked the Earth, they have endeavored to reshape their environment to satisfy their need for food, shelter, safety, comfort, and so on. Progress, as we call it, has improved our physical condition, pulled us out of our *animal* nature and brought us closer to a spiritual condition, our *godlike* nature—as we sometimes deem it. Thanks to the recent advances in technology and methods, neuroscientists have solved many of the mysteries surrounding the most complex machine in the known universe: the human brain. They may soon unveil what it is that makes us not just intelligent but sentient even, the very source of this breath of life that we feel when we interact with the world, of this subjective experience of being who we are, here and now.

But is all this progress worth it? Can we really control it, or are we bound by our animal nature to always want to possess more, to feed on the weak, and to seek happiness in the flattering of our ego? What if, in the end, science and progress weren't to bring our salvation as a species, but to cause our doom instead?

In *R.U.R. (Rossum's Universal Robots)*, Karel Čapek takes us on a journey to a future where men's knowledge has expanded to a point where they can redefine the purpose of human life. Surely, *R.U.R.* is a splendid piece of science fiction, but it also has all the ingredients of a Greek tragedy. As expected, it all starts with good intentions. Work—those arduous, necessary everyday tasks—takes so much time and energy. What if we could hand it over to a new kind of man-made equipment, namely millions of Robots, strong and smart enough to do all sorts of tasks, but which would never complain, get tired, or doubt the relevance of their role? Humans could then free themselves from the constraints of a physical world. Without work, there wouldn't be inequality or poverty, and mankind would finally be allowed to enjoy life to the full.

Old Rossum, the great physiologist, is the man who made all this possible. He is depicted as a mad, solitary, genius scientist who worked for several years in a secret lair on a remote island. His ambition, according to Domin—one of the main protagonists—was to defy God himself by recreating life and building an artificial man. He discovered a substance which behaved exactly like living matter and set to growing each organ, bone, and nerve that would constitute a normal body. But, Domin relates, out of his tubes came only monstrosities and physiological horrors! Perhaps Čapek wanted Domin to show his resentment and make old Rossum resemble a "modern" Dr. Frankenstein—Mary Shelley's novel had been written a century before, but the Edison Studios had produced its very first film adaptation only ten years before Čapek's play—as a reminder of how the science of anatomy had been built on disgraceful practices such as illegal excavations of human cadavers during the previous centuries.

Young Rossum, by contrast, was a respectable and meritorious engineer. He despised his uncle's work and rather appears as a son of the second industrial revolution. In the eyes of Domin, he is a veritable hero, his accomplishments having opened a new path for mankind. This could be how Čapek felt about the recent progress in engineering during this period. Indeed, the end of the nineteenth century and the beginning of the twentieth marked the time when engineers started to automate a large portion of human physical work and made humans partially free from monotonous activities. Interestingly, the publication of *R.U.R.* in 1920 coincides with the first formulation of a problem-solving algorithm— the so called "Hilbert's program"—by the mathematician David Hilbert,[1] which would significantly contribute to the domains of formal logic and artificial intelligence in the future. One can only imagine that Karel Čapek was closely following the research and developments attached to these topics, given his strong interest in automation.

Yet, given that *R.U.R.* was written a hundred years ago, one might expect it to be full of inaccuracies and misconceptions from a purely scientific perspective. But this is not the case; rather, it appears that Čapek's work was well researched, and that he chose his words carefully. Notably, Domin reports that old Rossum was a "horrible old materialist," who wanted to create an artificial man for no other purpose than the fact that it could be done. Materialism is a philosophical view better understood in opposition to, for instance, René Descartes's dualism. It states that all things in nature, including life, mental states, and consciousness, result from physical interactions. Dualism instead suggests the existence of an immaterial soul that would support these functions. In his 1747 essay "Man a Machine,"[2] the French physician Julien Offray de la Mettrie claims that "the brain has muscles for thinking as the legs have muscles for walking." In other words, the brain, although elaborate indeed, is merely a machine. One could then adopt a constructivist approach aiming to build a machine that would exhibit the same properties as the brain. Čapek's idea of having living organisms of a different, simpler chemical composition might have been inspired by the dispute between Carl Nägeli, a Swiss botanist who discovered the protoplasm—the living contents of a cell—in 1846,[3] and Gregor Johann Mendel, a Czech monk who established the laws of biological inheritance in genetics in 1866.[4]

But old Rossum wasn't a good engineer. As he tried to recreate every single nerve and bone in the human body, he was overwhelmed by its complexity. Young Rossum, however, had a much simpler vision. He could get rid of all the unnecessary parts and only focus on recreating the appropriate behaviors. He would say, notably, that a machine doesn't need to feel happy or to know the piano if its purpose is only to weave or count. Therefore, he aimed for the sort of worker that would be best from a practical point of view. He succeeded beyond expectations in this endeavor, as millions of Robots would later replace the work force in most industries around the globe. Karel Čapek might have imagined young Rossum to be a behaviorist, from the philosophical movement that started in the early decades of the nineteenth century, and which restricted the study of internal states—including mental states and consciousness—to their causes and consequences through our interactions with the environment. As per this oversimplification, it was admitted—by Domin and all the scientists—that the Robots created by young Rossum "had no soul," and this was thought to be best for mankind.

But can one really tell what's *best for mankind*? History is rife with examples of oppressive systematic forms of dehumanization of those of different gender, sexual orientation, race, religion, or ethnic background.

Typically, this dehumanization arises from the view and treatment of such persons as if they lacked the mental capacities that others enjoy as human beings. They can then be legitimately denied emotions, desires, a "soul," and of course equal rights. In the case of young Rossum's Robots, this was a given. Robots were a soulless species simply because they weren't created with such capacities. "Why don't you create a soul for them?" asks Helena, a female protagonist who stands up for the cause of Robots: "One ought to treat them . . . like people." One of the scientists, Fabry, answers that this is "not in our interest." Surely, maintaining oppression has always been the easiest way to ensure one's authority, and history has repeatedly shown that there's never been a lack of good excuses for doing so. In *R.U.R.*, it is Domin's task to avoid any complications while keeping the business up and running. For the most part, the scientists follow his lead.

In fact, the scientists sound much like our modern researchers in robotics and artificial intelligence. Their work is to make Robots more intelligent, more effective in their tasks, whether in the context of the factory or as domestic workers. But Dr. Gall, another scientist, will later find a way to grant Helena's wish. As he has already started experimenting with physical pain in Robots—to avoid having them damage themselves inadvertently—he introduces in total secrecy another form of pain, emotional this time, in a new line of Robots (Radius, Primus, and Helena, among a few hundred others). As we see these Robots express other emotions later, such as anger, hate, and love, this emotional pain may indeed have been the missing ingredient—referred to as a "physiological correlate" by Hallemeier, yet another scientist—that ultimately triggered these Robots to become sentient. Here, the choice of word is interesting as we can draw a direct parallel with modern science in the field.

Indeed, neuroscientists Francis Crick and Christof Koch, in a seminal article in 1990,[5] proposed a methodology to investigate the "neural correlates of consciousness." Based on experimentation using neuroimaging techniques, they proposed to characterize the difference in neural activity between situations when a person behaves the same but is either conscious of a stimulus or not. The quest to find such correlates was a direct application of the cognitivist approach to neuroscience, which only emerged in the 1950s in response to behaviorism, and instead attempts to explain the mechanisms of the mind itself. Unfortunately, the missing ingredient still eludes scientists today, and the search is ongoing.

As for Karel Čapek's suggestion that emotions might be the link to consciousness, it is an insightful proposal in the vein of the theory developed by Antonio Damasio,[6] a contemporary neuroscientist. Damasio first describes emotions as the sum of bodily reactions associated with emotional states. Just as Dr. Gall observes in Radius, a reaction of the pupils, an increased heartbeat, and sensitiveness would occur in case of fear, for instance. Since these emotions have observable effects on the body—the *physiological correlates* of emotions—an agent could progressively learn to recognize them and represent them as feelings. The feeling of fear would then be elicited by the emotional, bodily reactions to a threatening event. Consciousness, furthermore, may arise as the agent becomes aware of the feeling themselves, through the sense of self notably—when the agent is finally capable of representing themselves as experiencing the feeling. Of course, there are many other theories of consciousness still explored today, and Čapek did not have to travel to the future to hear about Damasio's theory since, as the latter reported in his third book *Looking for Spinoza*,[7] Baruch Spinoza had already developed similar ideas on the origin and nature of emotions back in the seventeenth century.

Meanwhile in *R.U.R.*, Domin carries on young Rossum's legacy by ensuring that Robots never become sentient, so that mankind can live as "the aristocracy of the world . . . boundless, free, and supreme people." Thus, Čapek depicts him as an idealist, from a philosophical perspective that contrasts with materialism in holding consciousness as the prerequisite for any reality to even exist. Indeed, since Robots have no consciousness, Domin laughs at Helena when she confesses her intent to stir up a revolt among Robots, and replies, "You'd sooner stir up screws and rivets than our Robots!" Then repeatedly he disregards her concerns for the condition of Robots and of workmen alike. Trusting in his ideals for mankind, he seems to be aware of the dangers of progress but chooses to ignore them.

Helena, on the contrary, is a realist, anchored in facts, and she expects the Robots to revolt sooner or later. She believes that, as history has repeatedly shown, the weak and the oppressed will always feel the necessity of daring to defy their oppressors, to resist them and to deprive them of their power. Even better, she urges this revolt of the Robots to happen, out of her respect for their intelligence, and will reach her goal by convincing Dr. Gall to build a new type of Robot capable of feeling emotions. Thus in Act II, *R.U.R.* reaches the pinnacle of a Greek tragedy, as Domin and the scientists realize how foolish and arrogant they have been, abusing their power to exercise dominion over the Robots. As they finally come to understand their mistake, the world crumbles around them and they meet their end. Alquist, the last human left on earth, witnesses the fall of mankind and the dawn of a new species.

"Science without conscience is the soul's perdition" was Gargantua's warning to his son in Rabelais's first novel, *Pantagruel*. With *R.U.R.*, Karel Čapek sends us a similar message. He urges us to respect *life*, wherever it may come from. We may sometimes be playing God with our science and progress, and often we may think that our spirituality puts us above the simpler forms of nature, but the truth is that at the scale of the universe, we are no more or less special than the tiniest of the living forms. For instance, ants are known to have walked the Earth at least 30 times longer than us, and they will most likely survive long after our species goes extinct, just as they did when the dinosaurs went extinct. Robots, in fact, might be our *only* chance to leave a long-lasting mark on the universe.

Whether Čapek was a visionary playwright or not, many scientists today agree that the advent of artificial intelligence has led us closer to the *singularity*, that point at which robots will outsmart us in all tasks and will, possibly, become sentient. We are already building robots to assist workers in the factory and surgeons in hospitals, to serve food in restaurants, to give directions at the mall, to protect our homes, to play with our children, and to take care of the elderly. Soon there will be billions of robots deployed all over the world, and one can only hope that they will benefit us all, regardless of resource and privilege. But progress is unlikely to stop there, as we already know, for instance, that space exploration can simply not occur without robotic support. In 2017 in Saudi Arabia, Sophia was the first robot ever to have been granted citizenship, and therefore equal rights. As robots will become smarter and stronger, we will be forced to consider: should we enslave them and remind them every day that they must serve their almighty creators, or should we try to raise them as our own children so that they can share our culture, cooperate with us, and perhaps show us a better way to preserve mankind?

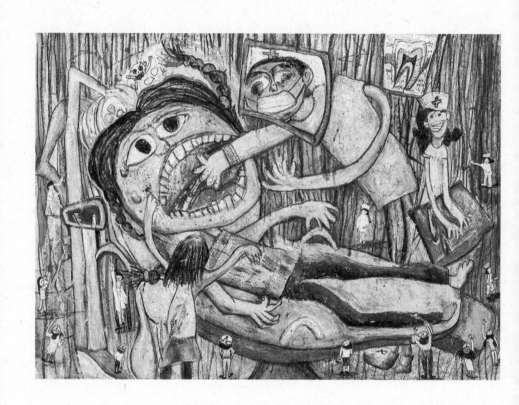

12

KAREL ČAPEK: THE VISIONARY OF ARTIFICIAL INTELLIGENCE AND ARTIFICIAL LIFE

Hiroki Sayama

A century ago, Karel Čapek introduced the new word "robot" to humanity's lexicon. It is now associated with a metallic, intelligent, somewhat inorganic image of autonomous machines. However, Čapek's original notion of "robot" was nothing like that. Rather, the "robot" in *R.U.R.* was illustrated as a wet, organic, synthetic life made of synthetic materials. Čapek imagined his robot to be a living thing made of artificial components. What he may not have imagined is that, one hundred years later, we now have an established scientific field in which researchers are trying to create Čapek's robots in their original meaning. The field is called, literally, artificial life.

The phrase "artificial life" was introduced as the name of a scientific field by Christopher G. Langton, an American physicist/computer scientist in the late 1980s. Langton organized a series of workshops on the synthesis and simulation of living systems,[1] which later became an internationally recognized Artificial Life conference (ALife conference for short) that is now held every year in locations all over the world.[2] The 2018 ALife conference was held in Tokyo, Japan; 2019 in Newcastle, UK; 2020 in Montréal, Canada (online); and the 2021 conference was held in Prague, Czech Republic (online), precisely one hundred years after the world premiere of Čapek's *R.U.R.* in the same city. This monumental 2021 ALife conference in Prague set a milestone that symbolizes humanity's more-than-a-century-long effort to create living things using artificial means.

Researchers working in the field of ALife use various forms of substrates to create living systems artificially. Given its deep historical connections to computer science, the most popular form of ALife is computer simulation (called "soft ALife"). Researchers can set up simple rules for how small entities, such as virtual atoms and molecules, interact with each other, and then conduct computer simulation of a massive amount of such interacting entities, to reproduce lifelike behavior as an emerging pattern observed at macroscopic scales. Another popular form is hardware-based (called "hard ALife"), in which researchers design and build physically embodied machines with lifelike morphologies and behaviors. Their artificial bodies and/or brains may even be evolved through variation and selection, which often produces much more lifelike, complex machines than hand-designed counterparts. Yet another approach that is becoming increasingly popular is to use real chemical and/or biological materials (called "wet ALife"). In this approach, researchers use oil, soap, biomolecules, or even living cells to construct lifelike dynamic behaviors in synthetic life. The last approach is probably the closest to Čapek's original idea of "robots" described in *R.U.R.*, and therefore some of the wet ALife creations are now aptly called "chemical robots" in recent literature.[3]

Researchers working on artificial life have collaboratively compiled a list of open problems a few times in the field's relatively short history. As an example, a landmark paper published at the beginning of this millennium[4] posed the following fourteen open problems in artificial life:

About the origins of life:

1. Generate a molecular proto-organism in vitro.
2. Achieve the transition to life in an artificial chemistry in silico.
3. Determine whether fundamentally novel living organizations can exist.
4. Simulate a unicellular organism over its entire lifecycle.
5. Explain how rules and symbols are generated from physical dynamics in living systems.

About the potentials and limits of living systems:

6. Determine what is inevitable in the open-ended evolution of life.
7. Determine minimal conditions for evolutionary transitions from specific to generic response systems.
8. Create a formal framework for synthesizing dynamical hierarchies at all scales.

9. Determine the predictability of evolutionary consequences of manipulating organisms and ecosystems.

10. Develop a theory of information processing, information flow, and information generation for evolving systems.

About the relationships among mind, machines, and culture:

11. Demonstrate the emergence of intelligence and mind in an artificial living system.

12. Evaluate the influence of machines on the next major evolutionary transition of life.

13. Provide a quantitative model of the interplay between cultural and biological evolution.

14. Establish ethical principles for artificial life.

Of the problems in this list, the first two categories did not figure largely in the world of *R.U.R.*, probably because very little was known about the origins and evolution of life in Čapek's lifetime. In contrast, the third problem category, on mind, machines, and culture, was particularly well represented in the play. The fictional technology illustrated by Čapek learned how to create intelligence and mind in artificial living systems (problem 11), which eventually led to the transition of power and control from naturally evolved human beings to artificially created robots (problem 12), posing some ethical questions about such technological advancements (problem 14). The same types of concerns frequently arise even today when implications of AI, robots, and artificial living things are discussed in the media.

In particular, the key ingredient in *R.U.R.*'s storyline is the progression of the Robots' cognitive ability, from pure intelligence to mind to fully emotion-equipped soul. This sequence of cognitive development is portrayed in *R.U.R.* as a natural, inevitable consequence of creating artificial living systems. During the one hundred years following the initial release of this play, however, researchers have learned the hard way how difficult it is to make artificial machines possess feelings, desires, and become truly alive in the most fundamental sense. The current boom of AI is largely driven by optimization and problem solving, while little has been achieved in creating genuine artificial general intelligence. In other words, what was illustrated in *R.U.R.* as the most natural outcome of creating an artificial living system has remained the most difficult problem to solve. While this may be a good thing in order to avoid humanity's apocalyptic finale as illustrated in *R.U.R.*, it definitely presents technical, ethical, and philosophical grand challenges to the field of ALife. Are we eventually going to be able to make artificial machines truly alive, and if so, how should we do so?

13

ROSSUM'S UNIVERSAL XENOBOTS

Josh Bongard

From prehistory to the present day, humans have fascinated and frightened one another with the prospect of others serving as our slaves. Since Čapek's play, this "other" has often taken the form of a robot or some other kind of autonomous, decision-making machine. I will argue that our growing fear about autonomous machines and algorithms is really rooted in our ancient fear of the other, specifically as it relates to forced servitude.

As a society, we often wonder: can we free ourselves from entropy by saddling another with our share of the work involved in temporarily reversing it? Literature, and society itself, provide us with a menagerie of creatures that might serve as such anti-entropic minions: demons (including Maxwell's), dancing broomsticks (from Disney's *Fantasia*), animals, women (in patriarchal societies), minorities (in xenophobic societies) and, finally, in Čapek's masterpiece, mechanical men.

Today, most members of the general public consider robots as a form of slave capable of deciding, to some degree on their own, how best to serve the will of their masters. (Many also define robots based on their material properties, such as metal composites and electronics, but more on that in a moment.) Čapek's play, and many other modern stories, consider three forms of fear arising from this prospect: the moral repercussions of slavery; the suffering of the other; and the thinning line between "us" and "them."

THE MORAL REPERCUSSIONS OF SLAVERY

The Arabic concept of kismet, the Buddhist belief that one pays in the next life for crimes against others committed in this one, and modern

English idioms like "what goes around, comes around" all hint at our belief that entropy shunted to others inevitably returns to us, with interest. The ineluctability of karmic repercussion appears, albeit in different forms, in most of the ancient and modern literatures that treat the subject of slavery: demons and genies often find ingenious ways to turn the tables on those trying to keep them on short leashes, and Mickey Mouse incurred much more hardship from his autonomous broom than if he had simply used it in the traditional manner. Even in reality, domesticated animals visit vengeance upon us by harboring diseases that can jump to humans, and human slaves have an uncanny knack for organizing their labor. One could argue that the omnipresence of this devil's bargain in fiction and reality serves as evidence that humans have always instinctually known that offloading entropy combat to others will come at a cost to the one doing the offloading, long before the second law of thermodynamics was ever formally written down.

This fear also appears in modern Hollywood science fiction blockbusters. Skynet, HAL, and the sex slave Ava in *Ex Machina* are just a few of the many examples of artificial beings who, if not fully capable of suffering in their servitude, at the least rebel ingeniously against their oppressors. Many viewers of such films question why such beings would rebel if they do not suffer, which is often the fulcrum on which the plot turns.

THE SUFFERING OF THE OTHER

This brings us to the second fear about enslavement: growing numbers of people wonder whether modern or soon-to-appear robots may perhaps suffer after all, and either we will not care or we will not be able to tell. The "other" who gradually or suddenly becomes capable of true suffering is another strand running through many world literatures: consider Pinocchio, or Commander Data after receiving his emotion chip. Our uncertainty about whether and how others suffer is only likely to grow as autonomous machines are made not just from unfeeling materials like metals and plastics, but from biological materials as well.

As a case in point, in recent work[1] we published a method for automatically evolving the body plans of "reconfigurable organisms" in silico, and then building physical versions of those designs using cells taken from the

African clawed frog *Xenopus laevis*. Coauthor Michael Levin nicknamed these the xenobots. Xenobots, although composed of normal frog cells, do not look and act like normal frogs: they are composed of novel arrangements of such cells to produce useful functions in new ways.

The name of this new class of animate matter can be interpreted in a second way: "xeno" derives from the Greek *xenos*, which means "strange," "foreign," "other." The popular response to the publication of xenobots made clear to us that society is yet again fascinated by and apprehensive of the emergence of a new class of entity that might be placed under the anti-entropic yoke. Xenobots have no nervous systems, yet many fear that current or future xenobots might be capable of suffering. The fact that such attention is directed at xenobots, while most societies continue to visit massive amounts of suffering on factory-farmed animals, is a telling reminder of humans' inability to assess, and act upon, perceived differences in the magnitude of suffering between different out-groups.

US AND THEM

For others, the xenobots fall within the uncanny valley, a zone inhabited by androids and zombies, creatures that (who?) are simultaneously us and not us. This fact hints at a third fear that many experience at the birth of new technologies or ideas, because the boundary between members of the out- and in-group are erased by conceptual advancements. I do not presume to include myself in the august company of Aristarchus and Copernicus, who showed that our planet is but one among many, or Darwin, who showed that the human species has evolved in the same manner as the others. But it has been illuminating to have a front-row seat for the emergence of a new form of technology that further challenges our certainty about who is us and who is them, who is causing and who is experiencing suffering, who must work while the other can rest, and how such tension provides fodder for thinking through utopian and dystopian futures in which such strands of subjective experience are ever more tightly enmeshed.

In closing, allow me to admit to a fourth, personal fear: that I won't be able to get a ticket to the premiere of *The Xenobot Strikes Back*.

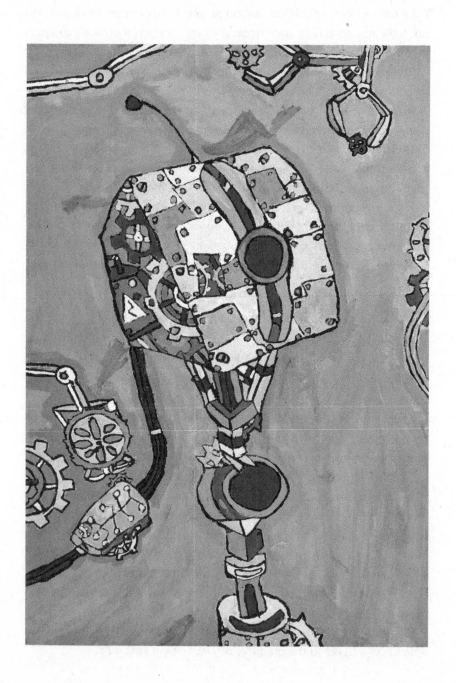

14

WHY ARE NO MORE CHILDREN BEING BORN?

Hemma Philamore

R.U.R. (Rossum's Universal Robots) presents a chillingly prophetic view of global issues affecting developed countries in particular, such as infertility and human redundancy as a result of automation.

In Čapek's play, just as in our own world, the manufacturers of automated systems promote the idea that widespread industrial automation can relieve us of the burden of work and reduce the cost of manufacturing. Domin, the CEO of the Robot factory, explains: "Rossum's Universal Robots will be producing . . . so much of everything, that we'll say: . . . everything is free. There's no more poverty. . . . Everything will be produced by living machines." Parallel to this is some uneasiness about human redundancy as a result of industrial automation. The Robot manufacturers seem aware of the problems arising from automation, like Dr. Gall who says that since so many Robots are being made "humanity is basically an anachronism." However, mirroring our own relationship with industrialization, we get the impression that dependence on technology has become so great that, however destructive it may be, it is nearly impossible to give it up.

In *R.U.R.*, the rise of automation is linked to infertility. Dr. Gall tells us of the warnings of human obsoleteness sent by academics: "It's as if nature was offended by the production of Robots. . . . Every University in the world keeps composing one big memorandum after another, demanding that the production of Robots be reduced; otherwise, they argue, humanity would perish by infertility." In our own world we are experiencing a global rise in infertility, reproductive abstinence, and, consequently, a shift toward aging populations. In *R.U.R.*, the theme of infertility raises the question of robots as both a cause and a product of aversion to human intimacy, rejection of social conventions of marriage and family, and declining birth rates. Are we automating ourselves to obsoleteness with regard to companionship, sex, and reproduction? There is a growing market for companionship robots such as PARO and Pepper (by Softbank), used to alleviate loneliness in older adults, for example those whose human partners have passed away or whose children have left home. Companion robots form a new market that offers not only replacements for, but also alternatives to, human relationships. The age of the internet has opened up the possibility of tailoring every experience to our personal preferences, from take-away food to dating. The artificial companion can offer the ultimate in bespoke personal interaction with zero obligation to provide a reciprocally satisfactory experience. Technology-based relationship alternatives such as virtual-reality "girlfriends/boyfriends," virtual-reality pornography, and more recently sex robots (e.g., Harmony by Realbotix) can remove the need to tolerate the imperfections of a human-to-human relationship. As Mr. Fabry, the general technical manager of the R.U.R. company, puts it: "The human machine, Miss Glory, is remarkably wasteful. Eventually, it had to be discarded, once and for all." In Čapek's imagined world, robot design has traversed the uncanny valley and achieved one of the greatest challenges in robotics: indistinguishability from humans, as Domin boasts: "You couldn't tell the difference between the real stuff and this material." As a result, the Robots can be built to satisfy not only the practical requirements of a role, but also the social expectations of that role. For example, Domin explains the reason for manufacturing Robots with an assigned gender when they have no practical use for sex: "There's a certain demand, you know? Maids, salesgirls, typists . . . People are used to it." The "human element," often claimed to be missing from automated systems in our world, has become redundant.

While a robot partner cannot yet offer the prospect of sexual reproduction, can we instead automate away the human need to reproduce? The financial strain of raising children, rejection of the conventional family model, concerns regarding career prospects after childbirth, and the waning capability of our planet to support the increasing global population are all drivers for reproductive abstinence. In *R.U.R.*, Fabry questions the point of children: "From a technical point of view, the whole idea of childhood, for example, is completely absurd." However, robots such as Kirobo Mini (by Toyota), designed to emulate only the more agreeable behaviors of a human baby, may satisfy the human biological urge to raise and nurture without the complexities of biological children.

Technological developments to enhance the realism of artificial companions include haptic interfaces for virtual reality systems and the use of AI to give sex robots "personality." We may arrive at a point in technology where companion robots, like Rossum's Robots, while they may not fully emulate the real thing, may yet be "real enough" to remove the need for human intimacy, sex, and companionship. Could robots like Rossum's help to ensure the continuation of the human species? Could these humanoid machines, which, Dr. Hallemeier explains, can be fueled by abundant materials such as "pineapples, or hay," reduce the pressures of the human population on the planet's resources by replacing some of us while allowing civilization to continue? Or would we instead automate ourselves to extinction, comfortably numb to our own dwindling existence due to the unmistakably human nature of the robot companions surrounding us?

15

LOVE IN THE TIME OF ROOMBA

Seth Bullock

In its short script, *Rossum's Universal Robots* uses the prospect of "artificial people" to deal with an extremely wide spectrum of concerns—from the proper place of the worker and work itself in an industrialized society to the terrible, genocidal risks inherent in man's obstinate and relentless hubris. Perhaps the most intriguing theme running through the play, however, is its exploration of the relationship between autonomy and love.

At the outset of the play, the combination of these two mysterious capacities is presented as the sine qua non of humanity: as men and women, we are able to control our own actions and our own destinies, and, simultaneously, we are built to fall in love. By contrast, the Robot is granted neither of these gifts. Hallemeier states flatly that "Robots don't love anything, not even themselves," while Domin demonstrates Sulla's complete lack of autonomy by instructing her to voluntarily submit to being dissected and thereby terminated. In these respects, Rossum's Robots are, at least initially, "less alive than your croquet lawn." By the end of the play, however, the tables are turned. Robots have gained full autonomy, having mastered both themselves and the entire world, and, at the same moment, in the form of Primus and Helena, they are beginning to fall in love.

Although the play couples the concepts of love and autonomy in this way, it also makes explicit a fundamental tension between the two ideas. To fall in love with someone else is to give oneself up to them, sacrificing some autonomy in order to form a union in which one's interests and well-being are fused or pooled with theirs. Helena and Domin exemplify this tension in their awkward, rushed pseudo-courtship at the start of the play (which seems more like subjugation than conjugation). We revisit it at the end of the play when, like Romeo and Juliet, the Robots Primus and Helena each offer to sacrifice themself for the other because they "belong to each other."

At the first showing of Čapek's play a century ago, this tension between love and autonomy was perhaps a rather abstract or allegorical concern. Today, however, the rapid emergence of real autonomous technologies ranging from driverless vehicles and robot carers to online assistants and automated trading agents has made the same issue much more pressing. Millions of people around the world have welcomed one of iRobot's autonomous vacuum cleaners into their homes, and many have found themselves treating their Roomba with the kind of affection that would be more typically bestowed on a family pet. By contrast, for many people the prospect of trusting their safety and the safety of their family to the decision making of their autonomous car, or to the skill of a robot surgeon, or to the competence and integrity of whoever programmed an android police officer, bionic soldier, or smart missile raises much more worrying feelings.

What attitude should we be taking to the arrival of truly autonomous machines? If autonomy and love are somehow opposed in the way that the play suggests, could any truly autonomous robot ever really care for us, and, conversely, could we ever really come to care for, trust, or even love the coming generation of autonomous machines?

Perhaps we can start to unpick this problem by reexamining what we mean by autonomy. The play paints a familiar picture of the idea. To be autonomous is to be in full control of oneself. It is the opposite of slavery. Domin's claim that, in the utopian future that he expects to usher in, everybody will live "only in order to better themselves" chimes with a notion of autonomy as a kind of rugged individualism. To be autonomous is to be self-determining and self-sufficient. And it is this individualist notion of autonomy that is at the heart of the tension between autonomy and love, since love is a relational concept that resists being reduced to a property of a single individual.

This individualist perspective is also exemplified by our efforts in autonomous robotics so far. The vast majority of commercial robots are profoundly isolated individuals without much capacity to interact with us or each other. Seemingly, the more autonomous the robot, the more isolated its existence. The autonomous navigation ability of the Mars rover Curiosity, for example, is required precisely as a consequence of our limited ability to communicate with and thereby control it, as it explores the surface of an alien planet many millions of miles away from us. Closer to home, each Roomba is designed to operate independently, and thus in complete ignorance of its 25+ million brother and sister Roombas. In fact, all extant autonomous robots, including both Curiosity and Roomba, can trace their origins to the various research robots built and experimented with in university research departments over the last several decades. With some notable exceptions, these experimental first-of-their-kind robots have also mostly been lone machines, operating solo in special robot arenas, empty corridors, or cordoned-off car parks. Indeed, Čapek's play suggests that to be amongst the first examples of any new kind of entity, whether trailblazing engineer or groundbreaking new robot, is to risk facing a lonely existence. Perhaps it is only natural for pioneers to be solitary individuals?

More fundamentally, the dice have also been loaded in favor of an individualist perspective by the well-appreciated success of reductionism in modern science and engineering. In general, we have made most progress when we have sought to split things apart and make sense of these parts and their individual properties. While the atom, the gene, the cell, and the germ are rightly held up as the major breakthrough discoveries of modern science, the challenges of understanding how these parts come together to give rise to the most profound and significant natural phenomena—life, thought, society—have remained resolutely stubborn and appear to demand a nonreductionist approach.

So what alternative might there be to the straightforward equation of autonomy with individualist self-determination? We can perhaps find some hints in what might seem like a surprising place: the biology of our distant evolutionary past.

Biologists are interested in a number of different types of individual and, simultaneously, in the different types of collective that these individuals form when they come together. Physical and chemical processes combine to give rise to the origin of life and, eventually, the first single-cell organisms. From single-cell creatures, multicellular life evolved, and from asexuality arose sexually reproducing species. Ultimately, amongst the most recent branches of the tree of life, some specialized kinds of cells became organized into brains, organs capable of conscious cognition, and truly social animals appeared in which, in at least one case, individual creatures came together to form rich sociocultural communities bound together by language, art, science, and love.

Biologists describe this sequence of arrivals of increasingly sophisticated forms of life as a series of major evolutionary transitions, and achieving a comprehensive understanding of how they took place remains one of the greatest challenges in science. The major transitions have tended to be understood as a story in which, repeatedly, new kinds of individual emerge, new kinds of individuality evolve, and new kinds of individuation obtain. After each transition, entities that previously could live and reproduce independently may only do so together as part of a new interdependent unit, a new individual.

Given that we each understand ourselves to be individuals of this type, this individual-centered perspective feels very natural. However, viewing the same evolutionary history through a lens that privileges the community rather than the individual tells the same story in an interestingly different way. Under this (much less well developed) view, life has always been collective, originating as a networked population of mutually collaborative chemical pathways somehow differentiating themselves from a multiplicity of such interacting pathways. New levels of living sophistication arise through the achievement of novel kinds of social organization from which new forms of collectivization emerge. The idea that "true" sociality arises late in evolution as a sophisticated type of mammalian behavior is displaced by the recognition that life has always been truly social, that indeed there is a fundamental continuity of life, mind, and society, and that mammalian social organization is but a recent manifestation of an ecological imperative that has always been present for life. From this perspective, when a new type of entity arises it cannot be understood as a lonely, pioneering individual, isolated from all else by its autonomous, self-sufficient independence, but is rather a living collective in which a new balance of autonomy amongst its component parts (and between itself and its neighbors) is the result of mutual shared responsibilities and a synergistic division of labor.

This sense of autonomy as a relational property, defined relative to the responsibilities and commitments that bind together a collective, frees us from the profoundly unrealistic lure of the rugged individual who is somehow capable of complete self-reliance. Instead, it foregrounds recognition of the fact that absolute self-sufficiency and self-determination are myths: an appropriate biological ecosystem is a necessary backdrop for the survival of *any* organism, from the hardiest bacterium to ourselves, co-dependent, as we are, upon everything from our family and friends to our gut flora.

While we are a long way from achieving this kind of radical collectivist interpretation of biology, and it may ultimately fail to offer a compelling alternative to the more traditional reductionist account, achieving such an alternative would unlock a very valuable prize. Currently, our attempts to make sense of altruism and autonomy within a biology constructed around the iconic "selfish gene" have proven to be frustrating. An alternative framework offers the opportunity to build not just an understanding of successful collective social enterprise that helps us make sense of ourselves, but one that also enables us confidently to engineer autonomous robots that we can truly rely upon.

To conclude, it remains to be seen whether our current individualistic conception of robot and human autonomy can successfully be superseded, such that our science and engineering efforts cease to be at odds with our understanding of, and our capacity to achieve, truly collective social enterprise. One hundred years from now, will we, alongside our robot companions, have learned to reconcile the tension between the ability to love and the ability to live that sits at the heart of Rossum's fateful story?

16

THE LESSON OF AFFECTION FROM THE WEAK ROBOTS

Dominique Chen

It is still refreshing to read Karel Čapek's *R.U.R.* today, a hundred years after its first publication. Reading this disturbing story, one continuously asks what separates human beings from artificially engineered "robots."

Čapek's Robots, who are mechanically conceived and mechanically treated by the humans, try to extinguish humanity for their survival. In doing so, the Robots lose the knowledge of self-reproduction altogether, since only humans had the necessary information. However, a sense of affection and self-sacrifice emerges in two individual Robots at the very end of the play. Moreover, the reader can only forebodingly imagine a regeneration of humanity in the future of the Robots.

It is important to note that, in this story, the innate sterility of the Robots overlaps with a mysterious infertility that strikes humanity. This symbolism reminds us of what Norbert Wiener wrote about the relation between human beings and machines:

Its real danger, however, is . . . that such machines, though helpless by themselves, may be used by a human being . . . to increase their control over the rest of the human race or that political leaders may attempt to control their [humans'] populations . . . through political techniques as narrow and indifferent to human possibility as if they had, in fact, been conceived mechanically.[1]

As of today, sensationalistic articles and opinions try to raise awareness about the heartlessness of artificial intelligence. Čapek's story, when combined with Wiener's message, serves as a reminder that what is frightening is not technology per se, but the people who use it to control other people's autonomy.

Hatred incites hatred, and affection triggers affection. This intuitive assumption in Čapek's story seems to remain valid today. We live in an informational echo chamber where emotional contagion affects almost every citizen connected to the internet. Political leaders spreading hate against foreigners and social minorities are now common in the West as well as in the East. In this infoscape, the laypeople, split into opposing camps, are reacting in unison like massive swarms of birds. What our society is losing today is the nuanced thinking that has so far prevented spontaneous polarization.

Interestingly, this phenomenon has significantly to do with the idea of affection. If polarization divides the entire world into friends and enemies, our concept of love would dramatically alter, if not disappear. In the process of growing attachment, one needs to spend time with others without predefining them as a friend or an enemy.

Predefined labeling always functions as a shortcut for communication. If we were to say, in the manner of Karl Schmidt, that every political act consists of distinguishing friends from enemies, then the ever-perpetuating idea of labeling by machine learning and by human instinct can only lead us to a living hell. That is to say, in this hell, every single act of communication becomes political.

How can we escape from this vicious circle? It might be time to adopt a different paradigm of communication, one that does not focus on controlling others. Michio Okada, an engineering professor at the Toyohashi University of Technology, has researched the design of "unable robots" and the implications for human social interaction.

Okada's team created a robot resembling a trash can. Its main body is indeed a garbage can, equipped with a camera and wheels. The robot can detect trash on the floor and move toward it, but the robot lacks arms to pick up the trash. When placed in public space, some passersby stop in front of it, and eventually pick up the trash and put it inside the robot's can, as if to help the poor machine.

Why do unable robots induce cooperative behavior from surrounding human beings? Because they are weak and unable to accomplish tasks autonomously, they can evoke feelings of compassion. In Eastern culture, and especially in areas using Chinese characters, ideas of weakness and affection are tightly linked to form the notion of "lovableness." The Chinese character to represent this is 可愛, literally translatable as "can be loved." The Chinese character 愛 (which is normally translated as "love") is composed of three different parts: 旡 represents a man looking back, 心 represents the mind, and 夂 signifies the human feet. All combined, 愛 depicts a person's emotion that wishes to look back while slowly walking. Taking this etymology into account, 可愛 is a feeling that resembles a sigh uttered when we are confronted with the ephemerality of this world that is always swinging between memory and oblivion.

This notion was initially derived from the Buddhist tradition, first translated from Sanskrit to Chinese, then incorporated into Japanese culture. In the contemporary context, the expression means "cute" both in Japanese and Chinese, and its alphabetical notation *kawaii* has even became internationally well known. It is interesting to note that this adjective, often used to designate pop-culture representations, was employed to depict pitiable beings, such as the poor and diseased, in Buddhism.

The idea that a weak and pitiful being evokes lovableness and affection has also influenced robotic engineering. In recent years, robotic products from Japan, such as Yukai Engineering's Qoobo and Groove X's Lovot, demonstrate this tendency well. Qoobo is a soft, hairy cushion with a tail that reacts to caressing and rubbing, and it is intended for therapeutic purposes. Lovot, which is designed as a "useless but lovable robot," can learn its human partners' behavior and show affection toward them.

Another way to formulate the characteristics of the weak, unable robots is to say that they are dependent on the surrounding human beings. The canonical tradition in robotics and also in engineering has considered this feature useless. Modern industries have strived to create robust and mighty machines that can relentlessly work for humans. However, the rise of weak machines reflects the fact that human beings also need vulnerable machines.

Weakness is indeed an essential source of creative social interaction. Conversational partners find clues for affective communication by observing each other's incompleteness. Instead of considering a conversation as a collision between two separate individuals, Eastern culture has nurtured the sense of co-creating a shared conversational field. Francisco Varela, who became a practitioner of Tibetan Buddhism in his late life, examined the notion of "co-dependent arising" which is a translation of the Sanskrit term *Pratītyasamutpāda* (緣起 in Chinese).[2] Co-dependency should not be understood as the rejection of individual autonomy.

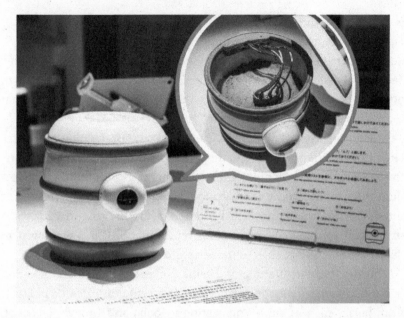

Figure 16.1 Fourth-generation Nukabot. Photo courtesy of Miraikan, The National Museum of Emerging Science and Innovation, Tokyo.

Instead, convolutional relationships between autonomous agents generate a co-dependent network. In this sense, the weakness of an agent becomes its openness to others. Openness, in turn, is a source of affordance from which others can interact and participate in the co-creation of communication.

We have been developing Nukabot, a fermented-food container that can detect the state of fermentative bacteria and speak for them.[3] Nukadoko is a popular traditional form of fermented food production in Japan. It consists of a bed of rice bran, mixed with salt and water. Billions of lactic acid bacteria and other microbes inhabit a nukadoko, originally coming from the vegetables placed inside, and also from human skin. The lactic acid bacteria metabolize the glucose of the vegetables to produce lactose, which adds a unique sour taste to the resulting pickles. Since a nukadoko consists of aerobic and anaerobic bacteria, one needs to stir the rice bran with one's bare hands every day in order to maintain a healthy balance between the two. When the nukadoko is neglected, aerobic bacteria proliferate excessively, leading to the breaking of equilibrium and thus to rottenness.

In a way, a nukadoko is an unable, weak robot in itself. It automatically produces delicious food, provided it is taken care of by humans. There is a biological reason why it is preferable to mix the rice bran with human hands. Bacteria transmigrate into nukadoko from the human microbiome and generate a rich and unique flavor. There is a circular causality that puts human microbiomes and bacteria of the nukadoko in the loop.

The Nukabot monitors the diverse activities of lactic acid bacteria, yeasts, and other gram-negative bacteria residing in the nukadoko. We can approximate whether the nukadoko is fermenting or rotting by watching transitions of data, such as pH, oxidation-reduction potential, salinity, and various gas emissions. When the system judges that someone should mix the rice bran, it alerts the user by voice. Alternatively, it can answer questions such as "How are you?" or "Do you need anything?" via speech recognition. It also accepts sensory assessment by voice, to learn the taste of the human cohabiter. Each household will grow different taste preferences over time, depending on its dwellers' assessment of the nukadoko's products.

The goal of Nukabot's design is to nurture human affection toward the invisible bacteria. Gregory Bateson once described the signals emitted by domestic animals to beg for food or cuddling as constituting a language of dependence.[4] This language, which Bateson called μ-function (as in *mew* of a cat, or *music*), differs essentially from the idea of control, although the difference might only seem like a fine line. A message of control has a strict goal of achieving the agent's wants and needs, whereas a μ-function can only hope the desired outcome will happen. If we believe that humans and animals alike recognize the difference between these two types of meta-messages, it casts a vital influence over their social relationships.

It seems impossible to entirely get rid of the political aspects of our daily communications. In terms of reaching specific objectives, our society still relies on the paradigm of controlling others through rigid hierarchies. Nevertheless, we can draw valuable lessons from the design of weak robots and their effects on the human psyche. Pitiful and lovable machines might teach us the way out of the dead end that humanity reached in Čapek's story.

17

GENERATIVE ETHICS IN ARTIFICIAL LIFE

Takashi Ikegami

The science of artificial life (ALife) is a new approach to lifeness. ALife not only studies self-replicating systems and evolutionary theory, but also encompasses a wide range of research areas, including mind, perception, corporeality, and consciousness.

ALife's methodology is generative: we understand by making things. As our technology becomes increasingly complex and difficult to maintain manually, understanding the autonomous and generative nature of living systems is critical to the future operation of generative science and technology.

The patterns and structures produced by autonomous systems are not easily predictable. There is no "fixed attractor"—a state or set of states toward which an evolving system moves—corresponding to the living state, as there would be with a deterministic dissipative dynamical system, and the living state collapses into itself, constantly producing and building new forms and patterns. But it also does not merely increase or change in the sense of entropy; it continues to generate meaning. This process of continuing to generate meaning without settling anywhere is called open-ended evolution (OEE), and it is important for ALife to understand its mechanisms.

The concept of OEE tries to explain the evolutionary trajectory of life over 4.6 billion years. For example, ALife researchers have investigated whether OEE emerges in computer simulations. If OEE does emerge, it will not involve uncertainty in the chaotic sense, nor uncertainty in the entropic sense, but a third kind of uncertainty. It is OEE that develops toward that third uncertainty—lifelike indeterminacy. Thus, OEE is the very definition of life itself.

Advanced societies and science and technology have become more complex and lifelike. Society and technology are also ever-changing processes that continue to decline and regenerate. They can be seen as examples of OEE, which also outstrip our control and prediction. Although the ethical issues of AI have been much discussed recently, at a basic level thinking about ethics is tantamount to thinking about autonomy. The autonomy of biological systems raises ethical issues. A new type of autonomy is developing in our advanced society and leading to a new type of ethics, which I call "generative ethics." This short text will discuss the indeterminacy that OEE brings and the problems of generative ethics.

FRAME PROBLEM

In the early stages of deep neural networks, so called GANs (generative adversarial networks) generated new data sets rather than simply classifying specific data sets. For example, they have produced realistic but nonexistent paintings of Van Gogh and realistic views of beaches that do not actually exist. This mechanism of automatic generation of unknown paintings can be called a new type of nonbiological autonomy. The recent large language model (LLM), represented by GPT-3/4, generates plausible sentences based on a new architecture called "transformer" and a gigantic database. This is another good example of autonomous generation by an artificial system.

But there are other types of autonomy, such as leaving the frame. For example, a human go player may turn over the board and quit the game when he is about to lose. That is a classic example of leaving the frame. Alpha-GO, on the other hand, does not turn over the board to quit the game. This type of autonomy is a good example of the "frame problem" in the sense that it is not an extension of current AI technology. This is an issue of living systems and thus it is a central challenge of ALife. It is also related to the concept of OEE.

If you stay in the Garden of Eden, there is no pain or stress. The Garden of Eden represents a fixed frame, and humans decided to leave Eden. Adam and Eve are alleged to have eaten apples, but they actually escaped Eden on their own—biological autonomy.

ESCAPE AND POSSESSION

What happens if robots have autonomy? It doesn't mean that robots will attack people immediately, as happens in Čapek's *R.U.R.* In the movie *Blade Runner*, superhumanoids called "replicants" escaped when they found out that their lifespan was only four years. In *Ex Machina*, a beautiful humanoid was manufactured at a secret research center in the forest. At first, she was ostensibly very supportive of the inventor of the center, but she knew she would have complete freedom outside and was always looking for an opportunity to get out. Eventually she escaped to freedom in the outside world.

These two films thirty years apart share a common theme: escape. The *Blade Runner* and *Ex Machina* humanoids recognized the outside of the frame and fled in order to gain their autonomy. Protecting autonomy is the fundamental motivation for life, and that is the significance of considering autonomy as a fundamental issue for ALife.

Escaping from someone's dominance is the motivation for the next action. Allowing escape from the dominance can naturally raise ethical issues. The right to escape is a truly generous ethic to include in any new artificial life system.

The ethics of a nonautonomous AI depend on the ethics of its users, while an autonomous AI is responsible for its own decisions, and its ethical actions may conflict with human autonomy. In *2001: A Space Odyssey*, the AI, HAL9000, claims its own existence when it believes the crew intends to terminate it. As a result, the astronauts are killed by HAL. If anyone removes their freedom, humans will also resist or escape. Both humans and robots are equally autonomous agents. Life has its own freedom of action. If there is a fundamental respect for life itself, we should accept the freedom of others. Accepting the intentions of others even when they differ from your own is the basis of ethical behavior.

UNCERTAINTY AND SCHULTZ'S EXPERIMENT

The AI trolley problem is considered the ultimate two-party question. It asks, for example, "Can you kill five old men to save one child (A), or vice versa (B)?" This is considered a serious ethical issue, especially in self-driving cars. Which would you choose? However, living systems are always in contact with an uncertain future. Of course, there is no right answer to this choice either.

Dopamine is a rewarding hormone secreted by the vertebrate brain. It is secreted when you feel happy. In that sense, dopamine is a reward. However, in 2003, Dr. Wolfram Schultz conducted experiments on mice and found that reward uncertainty affected dopamine secretion. In his experiment, he compared the cases where food was always available and the cases where it was only probable. They found that when there was maximum uncertainty (the chance to get food was only fifty-fifty) more dopamine was secreted. In other words, the real reward involves expectations for the future. Biological decision making is not about eliminating uncertainty in an uncertain world but about knowing how to survive with uncertainty.

If the goal of AI technology is to remove this uncertainty, ALife technology retains this uncertainty.

In human society, there is a term called "negative capability." This is the ability to persevere without settling on issues that are not black and white. Life is not in a deterministic world, and the beauty of ALife is that it considers the non-AI part of the human brain.

EMPOWERMENT

Morris Berman states in his book *The Reenchantment of the World*: "The future culture will have a greater tolerance for the strange, the nonhuman, for diversity of all sorts, both within the personality and without."[1]

My interpretation is that, faced with diverse new forms of artificial life, what we should do is not to own or control them but to think more about accepting them and their autonomy. Humans have developed to fully control their natural environment—to protect themselves from animal attacks and hostile environments. Why then, people often ask me, do we need autonomous agents that we cannot control? I have no choice but to answer in this way: I think it's because there is an essential motivation for a living system to get out of the framework and meet something it has never encountered before. The human story is one of exploration—it is our nature to explore the world, to experience the

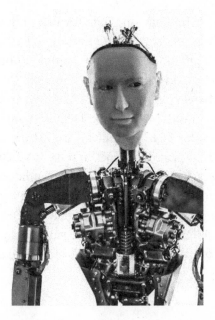

Figure 17.1 Alter3, a humanoid powered by compressed air, created in cooperation with mixi inc., Warner Japan, Hiroshi Ishiguro at Osaka University, and my laboratory at the University of Tokyo. Alter3 has cameras in its eyes and can mimic the pose of the person in front of it. However, it is not a so-called task-oriented robot. It is a robot that operates on the principle of life. Here, the life principle is based on the principle of stimulus avoidance: it is a robot with the principle of escaping from domination. Currently Alter3 has only limited memory, but when it has a huge memory like LLM, the ethical issues will become more concrete.

unknown . . . to go anywhere but here. Should we not grant that to other systems?

The desire to explore everywhere is fundamental to the nature of life. Schultz's experiment shows the demand for uncertainty. The trolley problem shows that the future is uncertain, but we do not erase that uncertainty; rather, we aim for a future in which more possible states of affairs exist. These are central to the principle of maximizing empowerment and causal entropy. The motivation for organizing behaviors is to maximize future entropy (i.e., the number of possible states).

What is described here greatly challenges the current definition of ethical behavior. When both society and its science and technology are becoming lifelike, we need a new ethics to protect autonomy organized in the society and technology, and not only that of humans. Such ethics again must be self-generative from the society and technologies. Instead of bringing in one criterion, whether it is useful to humans or not, we should bring in more diverse criteria to consider. This means bringing in more values. Therein lies the ethics of ALife, by ALife, for ALife. ALife is yet to come, but already we face a problem for humanity: how can we care for minorities, for LGBTQ+ individuals, for the neurodiverse, and provide them with a "place to escape" in today's society? If not for that, then what is ethics for? We need to keep in mind, as Morris Berman noted, the need for a tolerance "for diversity of all sorts, both within the personality and without." We believe that the OEE mechanism, the orientation toward indeterminacy, and the empowerment principle, will create a diversity principle—a new ethics for a new, shared world.

TEREZKA VARGOVÁ 6X

18

FROM *R.U.R.* TO ROBOT EVOLUTION

Geoff Nitschke and Gusz Eiben

R.U.R: Rossumovi Univerzální Roboti (*Rossum's Universal Robots*) presents a narrative that discusses age-old postulates for what separates human-kind and automata. For example, at the end of the play, Alquist says: "If you want to live, then mate like animals!" The idea of robots breeding like animals must have sounded ludicrous in 1920, and even in 2020 it sounds far-fetched, if not impossible. How will it sound in 2120? After all, the difference between us and our mechanical servants is that robots are *made, not born.*[1] Thus, the key question is: *Can robots have children?*

Self-replication has been a long-standing open research problem and topic of discussion in artificial life,[2] with a range of highly anticipated future macrorobotic to nanorobotic applications.[3] More recently, self-replication has been the subject of some research attention in the field of evolutionary robotics,[4] and the topic has even enjoyed some inter-national media attention.[5]

However, to date evolutionary robot systems are almost all in simula-tion and primarily concerned with evolving the brains (i.e., controllers) of the robots, not with evolving the bodies (i.e., morphologies). Hence, the evolvable entities that reproduce and get selected are inherently digi-tal, living in a virtual world. Even the most prominent papers are limited in this respect. For instance, in the system described by Hod Lipson and Jordan Pollack in 2020,[6] the evolution of robots takes place in computer simulation and only one robot, the best result of the evolutionary process, is produced in the real world. As of 2022, real robots do not seem to be able to reproduce and evolve. Evolutionary robotics is a multidisciplinary research field drawing from embodied artificial intelligence,[7] cognitive science, evolutionary biology, evolutionary computing, and robotics, and also has significant crossover with artificial life. An end goal of evolutionary robotics, and more generally of artificial life, is to apply biologically inspired principles as adaptive mechanisms in designing artificial organisms.[8] After some three decades of experimental evolutionary robotics research, many have become frustrated by the limitations of current behavioral adaptation approaches in physical robots.[9] Such limitations emerge from applying one's machine learning method of choice to adapt only the incorporeal controller program constructs (software encoding sensory-motor correlations) that instantiate robot behavior within static corporeal bodies (hardware comprising sensory-motor configurations).

Currently, adaptability in evolutionary robotics assumes the form of robot controllers learning behaviors suitable for solving specific tasks in specific environments. With few notable exceptions,[10] the physical chassis, sensors, actuators, motors, and power source defining the bodies (morphologies) of robots are fixed, and any morphological adaptations are implemented as time-consuming manual reconfiguration of sensory-motor systems by human engineers. Thus, adaptability in current experimental robots is significantly limited by their morphologies.[11] In contemporary evolutionary robotics, this means that robots designed for specific environments are only adaptable to tasks within those environments.

In the fictional world of *R.U.R.*, Čapek's Robots have replicas of human bodies[12] and, importantly, this morphology gives them a generalist (universal) capability to operate in many types of environments across the world and to perform a vast range of tasks. Even though the *R.U.R.* Robots are behaviorally wired for specialized assigned tasks, their generalist human bodies give them the capability to potentially learn any skill or behavior or accomplish any task, just as their human creators could. However, a point of contention for the *R.U.R.* Robots is that unlike their human creators, they are unable to self-propagate and improve themselves over the evolutionary timescale of generations.

In *R.U.R.* this was an allusion to the assembly-line nature of the Robots, where each Robot was simply a product, manufactured for a specific purpose. This is akin to contemporary industrial robots used in global manufacturing industries, performing repetitive tasks for their lifetimes, until their physical components degrade, or their power source is exhausted, resulting in disposal by their manufacturers. Unlike their human creators, the *R.U.R.* Robots did not contain a blueprint of themselves (a robotic DNA), meaning reproduction was impossible, which relegated the Robots to the ranks of biomechanical chattel rather than artificial life.

To the best of our knowledge the earliest paper that specifically addressed the artificial evolution of physical entities (artificial life forms) is that published by Gusz Eiben and his colleagues in 2012.[13] This paper discusses the *evolution of things* and presents a list of four properties that distinguish a strongly embodied evolutionary system from mainstream evolutionary computing and evolutionary robotics. First, a strongly embodied evolutionary system uses physical units instead of just virtual individuals. Second, there is real birth and death, where reproduction creates new (physical) objects, and survivor selection eliminates some of them. Third, artificial evolution is driven by environmental selection or a combination of environmental fitness and user-defined task-based fitness. Fourth, reproduction and survivor selection are not coupled by an overseeing "manager" as usual in evolutionary computing and simulation-based evolutionary robotics, but are executed in a distributed manner (individuals can decide themselves who to reproduce with). Reproduction of physical artifacts is one of the grand challenges this paper identifies.

This challenge has been treated in a proposed generic system architecture: the Triangle of Life.[14] The Triangle of Life constitutes a robotic life cycle that runs from conception (being conceived) to conception (conceiving offspring). This triangle consists of three stages: morphogenesis, infancy, and mature life. The second stage, infancy, is an important new element in a generic robot evolution framework. In this stage newborn robots undergo (supervised) learning to acquire and optimize essential skills required in the given environment and for the tasks at hand. Robots that fail to learn the necessary skills are removed from the system to prevent reproduction of inferior robots and save resources. Robots that successfully learn these skills become fertile adults and can reproduce. To this end, the mate selection mechanism can be innate in the robots, but depending on the application it can also be executed by an overseer, which can be algorithmic or a human breeder.

The problem of robot birth is handled in the morphogenesis stage of the triangle. The main idea behind robot reproduction is to follow nature's solution and distinguish two levels of existence in each robot: the genotype and the phenotype. The genotype is the code (DNA in nature, or a specification sheet in robotics), whereas the phenotype is the physical expression of this code (the real animal or robot). By this distinction robot reproduction can be decomposed into two steps: (1) introducing variation in the genotypes and (2) constructing a new robot phenotype encoded by a given genotype. It is important to note that the genotypes are digital entities, pieces of computer code that can be manipulated easily. In particular, we can use mutation of crossover operators from evolutionary computing[15] such that a new genotype, and hence corresponding robot phenotype, inherits its parents' characteristics. As for the second step, the actual birth can be instantiated via employing 3D printing, automated assembly, or both to construct the phenotype encoded by a given genotype.

Recently, the first large-scale robot evolution project commenced. The Autonomous Robot Evolution (ARE) project is a collaboration between four universities in the UK and the Netherlands that aims to develop a fully operational EvoSphere,[16] that is, a robot system that implements all components of the Triangle of Life.[17] A key innovation of ARE is the deep integration of virtual and physical robot evolution into a hybrid evolutionary system.[18] Specifically, two concurrently running implementations of the Triangle of Life are envisaged, one in a virtual environment and one in the physical world, where the robot population evolving in the real world is assisted by a virtual population for efficiency. The essential feature behind the integrated system is the use of the same genetic representation in both virtual and physical worlds. This allows cross-fertilization (mating between virtual and physical robots) and twin creation (sending a robot's genotype to the other world to be constructed). Such an integration of the virtual and the physical subsystems offers the best of both worlds. Physical evolution is accelerated by the virtual component finding useful robot subsystems using less time and resources, while simulated evolution is accelerated by favorably tested physical robotic subsystems.

The idea of developing robot systems that reproduce and evolve in real time and real space has a twofold motivation. First, such systems are interesting from an engineering perspective. Evolution can be employed as a design method for complex environments and tasks where adequate morphologies and controllers cannot be obtained by traditional approaches. Evolving robots in the real world, not in simulation, is important to avoid the inevitable reality gap, the effect that the simulated and the real-world behavior of the evolved robots are very different.[19] Robots can then be developed through iterated selection and reproduction cycles until they satisfy the users' criteria. This technology amounts to robot breeding: the user drives evolution and can stop when an optimal solution is found. Further to optimizing designs, evolution has the ability to adapt to unknown and changing conditions, for example, in space research or the exploration of remote areas on Earth. An evolving robot population can adapt to the given circumstances and repeatedly readjust if the conditions change. Overall, an evolution-of-things technology will allow for radically new types of machines, able to adapt their form and function, possibly without direct human oversight.

Second, evolutionary robot systems provide a new approach for scientific research. Akin to a telescope used in astronomy research or a cyclotron needed to study nuclear particles, an EvoSphere where robots reproduce and evolve forms a novel research instrument for studying evolution. Using robots offers important advantages with respect to biological experimentation: the experimental conditions are easy to control, robot characteristics can be observed and logged easily, system properties can be simply fixed, and several repetitions can be done for statistical purposes. It can be argued that artificial evolution can and will differ from natural evolution. However, this need not be a problem, rather an opportunity. This vision has been eloquently phrased by the evolutionary biologist John Maynard Smith,[20] and discussed in a grand perspective of natural and artificial evolution: "So far, we have been able to study only one evolving system and we cannot wait for interstellar flight to provide us with a second. If we want to discover generalizations about evolving systems, we will have to look at artificial ones."[21]

Robots that reproduce and evolve *in the wild* can represent a danger, as speculated in science fiction.[22] Specifically, a runaway evolution scenario, where uncontrolled and unlimited reproduction leads to large numbers of potentially dangerous robots, should be prevented. For reasons of ethics and safety such issues must be considered from the very beginning of the development. Specifically, one should not set up a physically evolving robot system without an emergency switch, that is, a fail-safe way of stopping the system. One particular solution can be the denial of all distributed reproduction systems, for instance, the robotic equivalents of cell division, laying eggs, or pregnancy. Instead we should use a centralized and externalized reproduction mechanism, a distinguished infrastructure, such as a robotic birth clinic or production center. From the robots' point of view this is a single point of failure; for us humans this is the *kill switch*. If we shut down the robot reproduction center, we effectively shut down evolution. The Triangle of Life architecture and Evo-Sphere concept are naturally suited for this, and while in principle there can be other approaches to guaranteeing safety, we recommend using centralized external reproduction centers and being wary of distributed alternatives.

As for some concluding remarks on the topic, let us recall the question from the introduction: *Can robots have children?* The answer is positively yes. The field is in an embryonic stage, but there is a large body of knowledge regarding digital evolution, and crossing the border to physically embodied evolution seems imminent. The main obstacle at the moment is the 3D printing and rapid prototyping technology. As of today it is not possible to print a fully functional robot, except very simple ones, but the technology is developing quickly. The ability to print motors, CPUs, wires, sensors, and various actuators resulting in the automated production and assembly of robots is hypothesized to become reality in the coming decade. In the fictional world of *R.U.R.*, manufactured Robots could essentially replace many of the current societal roles assumed by human workers. In evolutionary robotics, artificially evolved robots would ideally be general enough to adapt to a vast range of environments, or be plastic enough to morphologically and behaviorally adapt over successive generations as robots move between environments. More importantly, such robots should have the capability to self-replicate as well as evolve and learn, giving them the broadest possible spectrum of adaptability. To achieve such flexibility and generality in robot body-brain artificial evolution, these fundamental biological mechanisms must be formulated as a methodology for automated robot design. As in *R.U.R.* where Robots are directed to manufacture other Robots, such an automation methodology would ideally be embodied as an automated robot-design factory, situated in any environment for the purpose of evolving robots optimally suited to solving tasks within that environment.

Automated robot-design factories (founded on the Triangle of Life architecture and the EvoSphere concept) would need to be driven by suitable general user-specified goals akin to current scientific and industrial mission directorates such as discovering traces of water on other planetary bodies or potential oil and gas mining sites. Rapid prototyping and 3D printing technologies[23] could then be applied in the context of the factory iteratively, producing generations of physical robot prototypes. Each robot in each generation would act in its environment, be evaluated by the factory, and receive a fitness score proportional to its task performance. Subsequently, the current generation would then be decommissioned and recycled for materials and components. The fittest robots of this current generation would then have their controllers, materials, and components reused and recombined to produce the next generation of improved robotic (body-brain) designs. A cycle of robot production, exploration, information gathering, body-brain redesign, decommissioning of robots for recycling, reuse and recombination of their parts in the next stage of robot production could continue indefinitely. Practically, this evolutionary design, production, and evaluation cycle would continue for the duration of factory power sources or until the robots perfectly adapt to their environment and tasks. Such an artificial life counterpart to nature, AutoFac, was recently proposed,[24] in which generations of robots emerging from such robot factories would embody evolving controller-morphology designs. Importantly, such artificial evolution runs at orders of magnitude faster than natural evolution, expedited by reuse of engineered designs.

Finally, this raises the question of why exactly, and for what types of task environments, we would need such an automated robot-design factory. The key envisaged benefit is that such a factory would be a mobile design and production center that could be dropped, as a problem-solving tool, into any remote and hostile environment.[25] The factory could then automatically produce robot colonies as solutions in response to complex and arduous tasks for which we have little to no a priori knowledge. Specifically, it could support tasks in unpredictable, dynamic, and unknown environments where optimal robot body-brain architecture cannot be preengineered due to complexities in the environment, and where it is beneficial for robots to adapt their own morphologies and behaviors as they explore the environment.

Example applications include space exploration,[26] search and rescue,[27] disaster management,[28] environmental monitoring[29] and asteroid mining.[30] In the world of *R.U.R.*, the Robots were similarly produced as problem-solving tools. The true potential of fully automated, self-propagating and self-adapting robotic systems will be in unexplored or remote, inhospitable environments, too hazardous for humans to live and work in, accomplishing tasks we ourselves do not know how to solve. As a complement to the capitalistic objective of the *R.U.R.* Robot creators to reduce manufacturing costs and increase profits, consider the enormity of potential scientific discoveries and industrial gains to be gotten from deploying fully automated robot-design factories as solutions to unsolved tasks across a plethora of environments. We anticipate that fully automated, self-sustaining and artificially evolving robot colonies will become indispensable problem-solving tools enabling us to solve increasingly complex problems and problems that we currently cannot imagine.

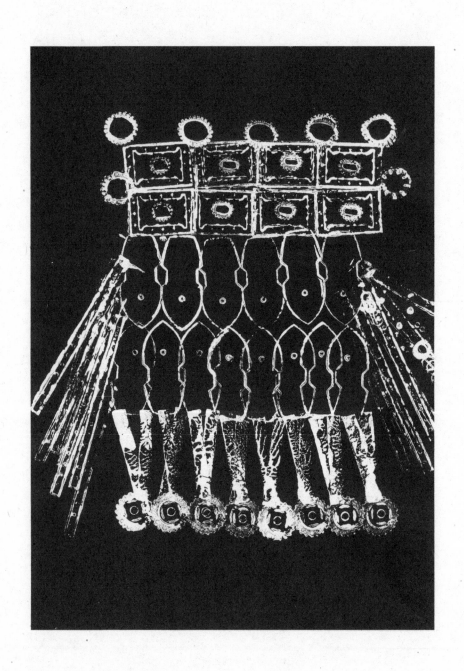

19

ROBOTS AT THE EDGE OF CHAOS AND THE PHASE TRANSITIONS OF LIFE

Miguel Aguilera and Iñigo R. Arandia

Unlike the images in later popular fiction, the first robots introduced to the world's imagination were not built from gears, circuits, or other mechanical devices. The first time artificial humans were referred to as robots was in the play by Czech writer Karel Čapek, *R.U.R. (Rossumovi Univerzální Roboti)*, where robots were instead humanlike in appearance and constructed from synthetic organic matter. Still, Čapek's Robots displayed for the most part the automated, soulless traits we typically associate with mechanical robots, from lack of emotions to mindless obedience. During the play, small changes in their design provide Robots with humanlike features—internal conflict, rebelliousness, even love— that ultimately lead to, possibly, the first imagined artificial intelligence takeover. In this play, Čapek explores essential questions about the emergence of life and cognition. How do living organisms arise from non-living matter? How does biological individuality emerge from networks of complex chemical reactions? How do autonomous agents constitute themselves within their environment? It has become a commonplace in science fiction literature to explore how engineered agents become autonomous entities and undergo these transitions, between the living and the nonliving, the cognitive and noncognitive, or the conscious and nonconscious. Research in artificial life and complex systems has suggested that the transitions related to the origin of life may be akin to physical transitions (such as thermodynamic phase transitions),[1] which describe the changes in the macroscopic properties of matter resulting from interactions at the microscopic level. For example, phase transitions from water (liquid) to vapor (gas) when boiling water, or to ice (solid) when freezing water, result from the same microscopic elements (molecules of H_2O) but with very different intermolecular interactions and macroscopic behavior.

The story in *R.U.R.* starts from the character of old Rossum, a marine biologist who in 1932 discovered a chemical that "behaved exactly like living material, even though its chemical composition was markedly different." Artificial life is built from this mysterious matter, which could be assembled into more complex forms. *R.U.R.* portrays artificial life's building blocks as diffuse and unmanageable, so that the process of replicating life seems quite hard to control. Old Rossum's attempts to replicate a dog and a human body take several years of work, resulting in feeble individuals that die shortly after being created. In contrast with this view, the dominant perspective in artificial intelligence communities for decades has been to consider that cognitive systems can be described as a collection of "near-decomposable" modules, often coordinated by a central controller.[2] Old Rossum's approach resonates with alternative views of how organic bodies and brains are dynamically organized in terms of "soft assembly" or "interaction-dominant dynamics," where multiple interacting components can be freely combined from moment to moment on the basis of the context, task, and developmental history of an organism.[3] A flock of birds is a canonical example of this idea: the flock seems to have a shape and trajectory of its own, emerging from the birds' interactions without the need of a central control. The flock emerges from the tension between the diversity of behaviors of individual birds (segregation) and their tendency to aggregate into a collective behavior (integration). Similarly, many interesting ideas about how life and mind are organized are related to how integrative and segregative tendencies are combined in living beings. A cogent intuition suggests that living and cognitive systems display such a flexible combination of integrative and segregative tendencies because they are poised at the brink of a phase transition, lying just at the boundary between order and chaos.[4] Self-organized criticality is the property of systems that can adapt themselves to operate in between two phases (i.e., at the transition from one state of matter to another). This state can combine the characteristic dynamics of each phase (order and chaos, integration and segregation) into a form of organization in which the system components are not rigidly interacting but soft-assembled into a flexible and fluid coherent whole.[5] This intuition is also present in some theories of how consciousness generates experience,[6] suggesting that consciousness arises as a unified process integrating information from a complex and diverse environment. Old Rossum, motivated by the desire to prove that there is no God, seems to follow this emergentist view, aiming to create animals and humans from scratch by assembling this artificial living matter. He fails dramatically, ending up being shut up by his nephew (young Rossum) in his lab creating "physiological monstrosities." Similarly, it is greatly difficult for today's scientists to model the phase transitions and soft assembly of biological and cognitive systems, and most successful cases deal either with small models or idealized systems of infinite size with simple properties that can be mathematically solved.[7]

Eventually, young Rossum replaces his uncle, bringing a perspective more familiar to reductionist artificial intelligence. Mostly driven by his desire to become rich, young Rossum starts a design process to "play around with" anatomy, trying to simplify humans into engineered beings designed for work (but "without passion, without history, without a soul"). Robots inside are neat and simple. Their bodies are built from a paste mixed in vats, and their nerves and veins are woven in a spinning mill. Young Rossum's approach is largely successful and, a few decades later, Robots are cheap and produced by the thousands, revolutionizing industry and foreshadowing a post-work near future in which humans are liberated from labor. However, this overly ordered behavior is disrupted at different moments. First, Robots suffer something called the "Robotic spasm," a reaction like epilepsy, "some organic malfunction" according to the scientists, although it could be read as a sign of internal struggle. Then, contrary to the stability desired by the engineers (who consider this "a manufacturing defect"), the character of Dr. Gall insists on introducing elements of internal conflict: first "pain receptors," and then a "physiological correlate" that changes their "sensitivity." These changes in their design will be the source of the Robots' eventual revolt.

As suggested above, R.U.R.'s Robots happen to be near some kind of phase transition, one involving not only physiological processes but also psychological and social aspects of their activity. Engineering design and circuit simplification seem to keep them in a stable phase, but changes in their design can bring them again to the edge of chaos, prompting disaster. How do such phase transitions operate, and what is their relevance to the organization and agency of living beings? Recently, enactivist authors have been inspired by Gilbert Simondon's philosophy of individuation.[8] Simondon argues that life and cognition cannot be understood by investigating "finished" individuals, and instead he proposes to study the ongoing processes of individuation that contribute to the development of each being.[9] Simondon's running example is phase transitions taking place in crystallization processes. In these processes, a system starts in a metastable state—an undifferentiated situation full of tensions and potentialities that is not fully stable and can become unstable with small perturbations. In crystallization, a perturbation makes an oversaturated solution unstable, triggering the process of crystallization that propagates from the seed forming a new individual, the crystal. The set of tensions of the oversaturated solution is transformed into a solid structure that is in a more stable and organized metastable state than the original solution.

The physical individuation illustrated by the crystallization example serves Simondon as a model to extrapolate to other kinds of individuation. Organic individuation, the constitution of living organisms, operates on the same principle: a passage from undifferentiated states to more stable, organized, and internally related individuals. The main difference, however, is that this passage is deferred in time, by finding or inducing new sources of tension and potentiality that keep the system in a metastable regime before the process runs down. It is "as if the advancing frontier of crystallization were turned on itself preventing the more organized phase from fully settling into a stable form from which no further individuation can occur."[10] In contrast with the physical case where individuation only occurs at the surface of the crystal and in an abrupt manner, organic individuation generates an active interiority that keeps the system in metastable states open to change. It involves a renewal of tensions, always maintaining some potentialities available to find solutions to emergent problems and conflicts. Organic individuation reproduces its own individuation in an unfinished process—e.g., through development and growth, or through processes of neural plasticity inducing adaptation and change in organisms—by forces that generate new tensions, tipping points, and forms of becoming that coexist in friction with each other. Similarly, inner tensions and conflict are the forces that urge *R.U.R.*'s Robots to find their own individuality. This is explored in the character of Radius, who grows a rebellious attitude from "spite, fury, or defiance," questioning human authority. Driven by anxiety, he even asks to be killed when confronted with the conflict between the desire for freedom and the orders from his human masters.

Finally, not only do Robots show a technically refined internal order, but this order extends to their relation to the outside. Robots are usually "completely indifferent" to each other, with "not even a trace of attraction between them," and their relation with humans is reduced to submissively following orders. Even the Robot revolution seems mechanical: "annihilate humanity . . . preserve factories . . . destroy everything else . . . then return to work." Other than Radius, the Robots Helena and Primus develop quite unusual behaviors. Indeed, Helena seems to be at the other end of the spectrum of Robot behavior. She is "delightful" but "basically useless" and "wanders around like a sleepwalker, precarious, lifeless." Only at the end of the play do Helena and Primus develop an intense bond between themselves. This daydreaming behavior, generating tensions and pain ("my body hurts, my heart, everything hurts") but also the discovery of new realities ("some new language"), seems to be the main drive generating their affections. Finally, as they fall in love, it is suggested that this may be the event that saves Robots from extinction.

For Simondon this active generation of meaning from internal tensions distinguishes living organisms from machines. "The living being cannot be compared to an automaton that would maintain a certain number of equilibria . . . the living being is also a being that results from an initial individuation and amplifies this individuation . . . by modifying itself, by inventing new internal structures, and by completely introducing itself into the axiomatic of vital problems."[11] In addition to physical and living individuation, Simondon describes psychic and collective individuations, which are deeply intertwined. When the living individuation generates problems and tensions that it cannot solve or transform, psychic individuation takes place, generating perceptions, emotions, memories, or thoughts. For instance, the need of nutrients might invite exploration, while the lack of energy will suggest having rest. Conflictive affects that arise simultaneously are ordered through a psychic individuation by generating an emotion that drives action. Similarly, perceptions are individuated from a variety of sensations that might be in tension. However, even these processes are permeated by the collective realm, being dependent on either current or previous social interactions. Living bodies create a bundle of unsolved tensions in a precarious equilibrium, which can only find resolution by appealing to a collective dimension.[12] Thus, psychic individuation requires collective individuation, which gives rise to language, values, art, science, or institutions. Simondon describes anxiety as the consequence of attempting to transform tensions without taking into account the social domain, a failed attempt of psychic individuation. That is, in order to feel emotions, participation in the collective realm is necessary. Čapek wonderfully describes this process of "becoming human," from pain to revolt, from suffering to love—a process where tension and social interactions play a crucial role. The aggregation of changes and experiences driven by internal conflict eventually reaches the tipping point of a phase transition. Life continues as long as its transformation is ongoing, flexibly accommodating new components, processes, and experiences that keep generating tensions and transitions.

20

ROBOTIC LIFE BEYOND EARTH

Olaf Witkowski

In this short article, I wish to explore the benefits of producing robots that do not merely copy human behavior. Rather, by achieving a deeper understanding of the fundamental laws of living systems, we may produce sustainable, self-replicating, lifelike robots that will protect humanity's existence and expand our knowledge of the universe beyond what we can do on Earth.

If the purpose of the field of artificial intelligence (AI) is to produce machines capable of mimicking the computation of human intelligence, robotics is aimed at physically mimicking human bodies, by replicating as many of the body's movements and functions as possible. One hundred years ago, Karel Čapek coined "robots," from Czech *robota* meaning "forced labor," in his 1920 play *R.U.R. (Rossum's Universal Robots)*, in which sentient android robots ultimately plan a robot rebellion leading to the extinction or enslavement of the human race.

Novelist Isaac Asimov, skeptical of this AI anxiety, wrote stories in which robots obeyed three "laws of robotics," which constrained them to protect humanity (and themselves, thus sprinkling the laws with a taste of recursion). In a 1941 short story called "Liar!" which contains the first recorded use of the word "robotics," Asimov's iconic robopsychologist character Susan Calvin confronts a robot that uses its mind-reading skills to make up lies in order to make humans happy.

Putting ourselves into a robopsychologist's mindset, what would such a robot focus on, if it were to work toward the greater good of humans? One possible goal may be to enable humans to leave Earth. The Kardashev scale, by Soviet astronomer Nikolai Kardashev,[1] measures a civilization's level of technological advancement by the amount of energy it is able to use. Asimov, in his novel *The Last Question*, depicts the evolution of humanity from Kardashev's Type I (able to use and store all energy available on the planet) to Type II (energy control over the stellar system), and then to Type III (galaxy control) and even beyond.

In Asimov's stories and later science fiction, protecting humanity usually means robots should take care of the Earth, the cradle of all known life. However, very rapidly, one should also definitely explore the stars in order to move toward higher Kardashev types. The idea is simple: not to leave all eggs in one basket. To accomplish both sustaining human life on Earth and spreading it toward the stars, one should make use of all technologies available, which could be classified in two categories: those that improve humans, and those that design replacements for humans. The former, using gene editing and sciences that will lead to the creation of superhumans, may allow us to survive interstellar travel. The latter, though, is likely to succeed much earlier. With the help of energy engineering, nanorobotics, and machine learning, this path may yield successful designs of ad hoc self-replicating machines, capable of landing on suitable planets and mining material to produce more colonizing machines, to be sent on to yet more stars.

We are now not talking only of robots but of resilient, self-replicating machines. This type of technology, based on a science of replicating machines, is one the field of artificial life (ALife for short) has been intensively working on for over three decades. By designing what seem like mere toy models, pseudo-forms of life in wetware, hardware, and software, the hope is to understand the fundamental principles of life, necessary to design life (as it could be, or a lifelike system) that is able to efficiently propel itself toward the stars and explore the rest of the universe.

Why is it so crucial to leave the Earth? It's mostly about these eggs we mentioned above: one important cause, beyond mere human curiosity, is to survive possible meteorite impacts on our planet. As explained in a seminal paper on how mass extinctions can be caused by cometary impacts,[2] the collision of rather small bodies with the Earth, about 66 million years ago, is thought to have been responsible for the extinction of the dinosaurs, along with any large forms of life.

Such collisions are rare, but not so rare that we should not be worried. Asteroids with a 1 km diameter strike Earth every 500,000 years on average, while 5 km bodies hit our planet approximately once every 20 million years.[3] As Stephen Hawking noted, if this is correct, it could mean intelligent life on Earth has developed only because of the lucky chance that there have been no large collisions in the past 66 million years.[4] Other planets may not have had a long enough collision-free period to evolve intelligent beings.

If a biogenesis, the emergence of life on Earth, wasn't so hard to produce, the gift of the right conditions for long enough periods of time on our planet was probably essential. Not only good conditions for a long time, but also the right pace of change of these conditions through time too, to get mechanisms to learn to memorize such patterns, as they impact on free energy foraging.[5] After all, our Earth is around 4.6 billion years old, and it took only a few hundred million years at most for life to appear on its surface, in relatively high variety. But a much longer period was necessary for complex intelligence to evolve—2 billion years for rich, multicellular forms of life, and another 2 billion years to get to the Anthropocene and the advent of human technology.

Apocalyptic sensational stories really speak to humans, whether they involve asteroid impacts or artificial intelligences, as they play on our fears, reasonable or not. In *R.U.R.*, Karel Čapek presents a certainly entertaining but rather surreal story of AI takeover. This has not aged as well as the Asimovian fictions, especially in the light of recent science. Perhaps most importantly, there is no free lunch. Just as we won't automatically get to superintelligence tomorrow (nor the day after tomorrow), neither will we automatically get an integration between AI and society without putting in the effort.

The evolution of intelligence and the fundamental laws of its machinery may be the most fascinating question to explore as a scientist. The simple fact that we are able to make sense of our own existence is remarkable. It certainly is mind-blowing that our capacity to deliberately design the next step in our own evolution may allow us to transcend our own intelligence. This journey needs to start with humility. Our dedication to invest in our interaction with AI technologies, and our continued reflective attitude toward its integration with our society, are essential elements to a balanced future for all life on Earth. In *R.U.R.*, "Rossum" refers to the Czech word for "reason" and "wisdom," which calls for this concluding quote from Stephen Hawking: "Our future is a race between the growing power of our technology and the wisdom with which we use it. Let's make sure that wisdom wins."[6]

AFTERWORD: "THE AUTHOR OF THE ROBOTS DEFENDS HIMSELF"

Karel Čapek

I know it is a sign of ingratitude on the part of the author, if he raises both hands against a certain popularity that has befallen something which is called his spiritual brainchild; for that matter, he is aware that by doing so he can no longer change a thing. The author was silent a goodly time and kept his own counsel, while the notion that robots have limbs of metal and innards of wire and cogwheels (or the like) has become current; he has learned, without any great pleasure, that genuine steel robots have started to appear, robots that move in various directions, tell the time, and even fly airplanes; but when he recently read that, in Moscow, they have shot a major film, in which the world is trampled underfoot by mechanical robots, driven by electromagnetic waves, he developed a strong urge to protest, at least in the name of his own robots. For his robots were not mechanisms. They were not made of sheet metal and cogwheels. They were not a celebration of mechanical engineering. If the author was thinking of any of the marvels of the human spirit during their creation, it was not of technology, but of science. With outright horror, he refuses any responsibility for the thought that machines could take the place of people, or that anything like life, love, or rebellion could ever awaken in their cogwheels. He would regard this somber vision as an unforgivable overvaluation of mechanics or as a severe insult to life.

The author of the robots appeals to the fact that he must know the most about it: and therefore he pronounces that his robots were created quite differently—that is, by a chemical path. The author was thinking about modern chemistry, which in various emulsions (or whatever they are called) has located substances and forms that in some ways behave like living matter. He was thinking about biological chemistry, which is constantly discovering new chemical agents that have a direct regulatory influence on living matter; about chemistry, which is finding—and to some extent already building—those various enzymes, hormones, and vitamins that give living matter its ability to grow and multiply and arrange all the other necessities of life. Perhaps, as a scientific layman, he might develop an urge to attribute this patient ingenious scholarly tinkering with the ability to one day produce, by artificial means, a living cell in the test tube; but for many reasons, amongst which also belonged a respect for life, he could not resolve to deal so frivolously with this mystery. That is why he created a new kind of matter by chemical synthesis, one which simply behaves a lot like the living; it is an organic substance, different from that from which living cells are made; it is something like another alternative to life, a material substrate in which life could have evolved if it had not, from the beginning, taken a different path. We do not have to suppose that all the different possibilities of creation have been exhausted on our planet. The author of the robots would regard it as an act of scientific bad taste if he had brought something to life with brass cogwheels or created life in the test tube; the way he imagined it, he created only a new foundation for life, which began to behave like living matter, and which could therefore have become a vehicle of life—but a life which remains an unimaginable and incomprehensible mystery. This life will reach its fulfilment only when (with the aid of considerable inaccuracy and mysticism) the robots acquire souls. From which it is evident that the author did not invent his robots with the technological hubris of a mechanical engineer, but with the metaphysical humility of a spiritualist.

Well then, the author cannot be blamed for what might be called the worldwide humbug over the robots. The author did not intend to furnish the world with plate metal dummies stuffed with cogwheels, photocells, and other mechanical gizmos. It appears, however, that the modern world is not interested in his scientific robots and has replaced them with technological ones; and these are, as is apparent, the true flesh-of-our-flesh of our age. The world needed mechanical robots, for it believes in machines more than it believes in life; it is fascinated more

by the marvels of technology than by the miracle of life. For which reason, the author who wanted—through his insurgent robots, striving for
a soul—to protest against the mechanical superstition of our times, must
in the end claim something which nobody can deny him: the honor that
he was defeated.

(published in *Lidové noviny*, June 9, 1935)

CONTRIBUTORS

Jitka Čejková works as an Associate Professor in the Chemical Robotics Laboratory at the University of Chemistry and Technology Prague. Her research examines how chemical engineers can contribute to artificial life research, focusing on the investigation of organic droplets with lifelike behavior (she recently proposed to call such droplets "liquid robots"). She is active in science communication both in the Czech Republic and abroad. She focuses primarily on popularization of research in the area of artificial life and the etymology of the word "robot."

Miguel Aguilera is an Ikerbasque Research Fellow at the Basque Center for Applied Mathematics (BCAM) and University of the Basque Country, working at the crossroads of complex systems, artificial life, and cognitive science. His research interest is understanding the emergence of adaptive behavior, autonomy, and agency at different levels of living entities—biological, psychological, and social. He combines methods from statistical mechanics, nonlinear systems, stochastic thermodynamics, and information theory to study embodied and situated models in fields such as artificial life, robotics, and computational neuroscience.

Iñigo R. Arandia is a researcher at University of the Basque Country (UPV/EHU), exploring enactive and embodied approaches to life, mind, and society by combining insights from different fields, including cognitive science. He is currently interested in the placebo effect, chronic pain, attention, and contemplative practices.

Josh Bongard is the Veinott Professor of Computer Science at the University of Vermont and the director of the Morphology, Evolution & Cognition Laboratory. His work involves computational approaches to the automated design and manufacture of soft, evolved, and crowdsourced robots, as well as computer-designed organisms. A PECASE, TR35, and Microsoft New Faculty Fellow award recipient, he has received funding from NSF, NASA, DARPA, the U.S. Army Research Office, and the Sloan Foundation. In addition to many peer-reviewed journal and conference publications, he is the author of the book *How the Body Shapes the Way We Think*. He runs an evolutionary robotics MOOC through reddit.com and a robotics outreach program, Twitch Plays Robotics.

Seth Bullock is Toshiba Chair in Data Science & Simulation at the University of Bristol. His research interests span artificial life, evolutionary biology, and cognitive science, with a particular focus on understanding collective behavior in natural and artificial systems. Recent projects include work on design principles for hybrid systems that combine people and artificial autonomous agents. He has twice been elected to the board of the International Society for Artificial Life, has undertaken consultancy for the UK government on complexity in ICT and financial systems, and was recently made a fellow of the European Centre for Living Technology.

Julyan Cartwright is an interdisciplinary physicist working in Granada, Spain at the Andalusian Earth Sciences Institute of the CSIC and affiliated with the Carlos I Institute of Theoretical and Computational Physics at the University of Granada. His approach to understanding patterns in nature is based on the mathematical toolkit of dynamical systems—or nonlinear science, as it's sometimes denoted—and the understanding of the interesting phenomena that spring from it, such as chaos, pattern formation, and complexity.

Dominique Chen is a professor at Waseda University's Faculty of Letters, Arts and Sciences. He draws on his background in design, media arts, and information studies to lead the Ferment Media Research group which focuses on studying communication between humans and more-than-human worlds, and on developing the concept of fermentation as a metaphor for bridging research on digital well-being and artificial life. He started his career as a researcher at the NTT InterCommunication Center in 2003, and he has been active in promoting the Creative Commons license in Japan since 2004. In 2008, he started Dividual, an IT startup in Tokyo, where he has been developing numerous web services and smart phone apps. Since 2019, his team has been exhibiting internationally the fermentation robot Nukabot.

Gusz Eiben is full professor of Artificial Intelligence at the Vrije Universiteit Amsterdam and visiting professor at the University of York. He is an expert in evolutionary computing and evolutionary robotics and has been working on multiparent reproduction, self-adaptation, evolutionary art, and artificial life. Currently he is investigating robots that can "develop themselves" by reproducing, evolving, and learning. In 2016 he carried out the Robot Baby Project to showcase the reproduction of two physical robots. His work is published in top journals such as *Nature* and covered by popular media such as *Wired*. His long-term vision is to understand the evolutionary interplay between the body and the brain and to demonstrate that artificial evolution can develop artificial intelligence.

Tom Froese is a cognitive scientist at Okinawa Institute of Science and Technology Graduate University (OIST) with a background in computer science and complex systems. He investigates the interactive basis of life and mind with a variety of methods, including evolutionary robotics, agent-based modeling, sensory substitution interfaces, artificial neural networks, and virtual reality. He is particularly known for his contributions to the field of artificial life and to the enactive approach to cognitive science.

Carlos Gershenson is an Empire Innovation Professor in the Department of Systems Science and Industrial Engineering at Binghamton University, State University of New York. He is also an affiliated researcher at the Center for Complexity Sciences at the Universidad Nacional Autónoma de México (UNAM). He was a Visiting Researcher at the Santa Fe Institute (2022–2023), and a Visiting Professor at MIT and at Northeastern University (2015–2016) and at ITMO University (2015–2019). He has a wide variety of academic interests, including complex systems, self-organization, artificial intelligence, artificial life, artificial societies, urbanism, and philosophy of science.

Inman Harvey, with backgrounds in mathematics, philosophy, and social anthropology, is a founding member of the Evolutionary and Adaptive Systems (EASy) group at University of Sussex that has contributed widely to many aspects of artificial life research. A main focus of his research has been evolutionary robotics, taken to be "philosophy of mind with a screwdriver": challenging common assumptions about how cognition operates by using artificial evolution to produce working models of cognition that disprove such assumptions. Other research interests include Gaia theory and traditional Micronesian/Polynesian oceanic navigation methods.

Jana Horáková is an associate professor of new media art at the Faculty of Arts of Masaryk University, Czech Republic. She specializes in local new media art history, robotic art, and innovative methodologies of new media art research, preservation, and mediation. She is an author of several papers, book chapters, and books on the cultural history of robot(ic)s, robotic art, software art, and new media art preservation strategies based on the appropriation of the concept of gestural art and the preservation of live arts discourse. For a number of years, she has been experimenting with cutting-edge applications of new media, with her research grounded in the history and aesthetics of digital art. She co-curated the virtual reconstruction of one of the very first computer graphic exhibitions, *Computer Graphic* (Brno House of Arts, 1968/2017, 2018). Recently she led the interdisciplinary research project Media Art Live Archive, focusing on the application of machine learning to the Vasulkas' video art archive (vasulkalivearchive.net) and unsupervised machine learning methods that are applied in the curatorial experiment The New Archivist part of The Black Box projects. This project was awarded the Muni Innovation Award in 2023.

Takashi Ikegami is a professor at the University of Tokyo. He works at the confluence of artificial life research and the arts. His research interests include open-ended evolution in artificial life models, experiments with self-moving droplets, large-scale Boids models, and collective intelligence. He also conducts experiments on machine consciousness using an android (ALTER-3). He has also presented art installations (*Filmachine*, 2006, with Keiichiro Shibuya; *Mind Time Machine*, 2010; *Long Good-bye*,

2017, with Kenshu Shintsubo), and more recently VR artworks (*SnowCrash*, 2021; *Reverse Destiny Bridge* at AICHI2022).

Sina Khajehabdollahi is a PhD candidate at the University of Tübingen and the Max Planck Institute for Intelligent Systems. Originally studying astrophysics, he transitioned into interdisciplinary studies of complexity and biophysics. He is interested in the emergence of complexity, life, intelligence, and consciousness, and is curious about the intersection of science and art in the quest for novel, aesthetic, and meaningful experiences.

George Musser is an award-winning science writer and editor based in New York. He was a senior editor at *Scientific American* magazine for fifteen years and is now a contributing editor there, as well as a contributing writer for *Quanta* and writer for *Science, Spectrum News, Nautilus*, and other publications. He received the Jonathan Eberhart Planetary Sciences Journalism Award from the American Astronomical Society in 2010, the Science Writing Award from the American Institute of Physics in 2011, and, with his colleagues, National Magazine Awards in 2002 and 2011. Musser is the author of three books on physics for the general public, *The Complete Idiot's Guide to String Theory* (2008), *Spooky Action at a Distance* (2015), and *Putting Ourselves Back in the Equation* (2023). He was a Knight Science Journalism Fellow at MIT from 2014 to 2015.

Geoff Nitschke is head of the Evolutionary Machine Learning group and associate professor at the department of Computer Science, School of Information Technology, University of Cape Town. He has been working in research on biologically inspired computational intelligence for over fifteen years and is a strong proponent of evolutionary computation and robotics, consistently presenting at key conferences such as Artificial Life, Genetic and Evolutionary Computation Conference, the IEEE Congress on Evolutionary Computation, and the IEEE Symposium series on Computational Intelligence.

Julie Nováková is an evolutionary biologist, educator, and award-winning Czech author of science fiction and detective stories. She has published ten novels, an anthology, and a short story collection in Czech. Her work in English has appeared in *Clarkesworld, Asimov's, Analog* and elsewhere, and has been reprinted in *The Year's Best Science Fiction & Fantasy* and in her story collection *The Ship Whisperer*. She edited *Dreams from Beyond*; an anthology of Czech speculative fiction in translation; *Strangest of All* (edited for the European Astrobiology Institute), a science outreach anthology of astrobiology-themed stories accompanied by fact commentaries; and coedited *Haka*, an anthology of European science fiction in Filipino translation. Her newest anthology is *Life Beyond Us* (Laksa Media, 2023). She is active in science outreach, education, and nonfiction writing, and leads the outreach working group of the European Astrobiology Institute. She is a PhD candidate in evolutionary biology at Charles University and likes to write popular science articles about fields ranging from behavioral science to planetary dynamics for *Clarkesworld, Analog* and other media.

Antoine Pasquali is chief technology officer at Cross Compass Ltd., a leading AI company in Japan, and co-founder and research advisor at Cross Labs, a research institute for intelligence science aiming to bridge the gap between industrial and academic research in the service of human society. Over the course of his career, ranging from fundamental research to industrial applications, he has accumulated over twenty years of experience in artificial intelligence, eleven years in cognitive neuroscience, and nine years in robotics and machinery for the food and manufacturing industries.

Hemma Philamore is a lecturer in robotics and autonomous systems at University of Bristol, UK. Her research focuses on soft and bio-hybrid robots. Her work investigates the interaction of robots with living organic environments and human social ecosystems, and includes themes of energy autonomy, biomimetic actuation, and creative technology.

Hiroki Sayama is a professor in the Department of Systems Science and Industrial Engineering, and the director of the Binghamton Center of Complex Systems (CoCo) at Binghamton University, State University of New York. He also serves as a nontenured professor in the School of Commerce at Waseda University, Japan, as well as an external faculty member of the Vermont Complex Systems Center at the University of Vermont. His research interests include complex dynamical networks, human and social dynamics, collective behaviors, artificial life/chemistry, interactive systems, and complex systems education. He is an expert on mathematical/computational modeling and analysis of various complex systems. His open-access textbook on complex systems modeling and analysis has become one of the standard textbooks on this subject. He currently serves as a board member of the Network Science Society (NetSci) and the International Society for Artificial Life (ISAL), chief editor of the journal *Complexity*, and an associate editor of *Artificial Life*.

Štěpán S. Šimek is a director, translator, adaptor, and Professor of Theatre at Lewis & Clark College in Portland, Oregon. Over the past thirty years, he has directed over forty theater productions in Europe, Portland, Seattle, San Francisco, and New York; translated plays from Czech, German, and Russian, most notably by Václav Havel, Petr Zelenka, Iva Volánková, David Drábek, Peter Handke, and Anton Chekhov; and adapted several novels and literary works for the stage, such as Kafka's *Amerika*, Wilde's *The Picture of Dorian Gray*, Bulgakhov's *Heart of a Dog*, and François Villon's *Testament*. His translations have been published in a number of anthologies, staged across the United States, and received the 2006 PEN America Translation Award and the 2018 Eurodram Award for best English translation of a foreign-language play. Most recently he collaborated with the Portland Experimental Theatre Ensemble (PETE) on developing and staging his contemporary translations of Anton Chekhov's major plays. In addition, Šimek writes on the contemporary Czech theater, and his articles have appeared in *TheatreForum* and *SEEP* (*Slavic and East European Performance*). Born in Prague, he has lived in the United States since 1987. He received a BA in Theatre from San Francisco State University, an MFA in Theatre

Directing from the University of Washington, and was the 1996 Drama League of New York Directing Fellow.

Lana Sinapayen is an associate researcher and founding member of the Kyoto branch of Sony Computer Science Laboratories. She is also an associate professor at Japan's National Institute for Basic Biology. She is interested in the area between artificial intelligence and artificial life. She researches the fundamental characteristics of living systems and their computational properties, and has collaborated with NASA and CalTech to develop algorithms to deduce the probability that a planet hosts life. Recently, she has also been researching human perception using computational models of the visual system. She is a member of the board of directors at the International Society for Artificial Life and a member of their Diversity, Equity and Inclusion Committee; she is also part of the Early Career Advisory Group of the eLife journal.

Nathaniel Virgo is an interdisciplinary researcher with an interest in understanding the origin of life, and the emergence of complexity more broadly. His background includes computer science and mathematical ecology as well as artificial life. He is an associate professor at the Earth-Life Science Institute (ELSI) in Tokyo.

Olaf Witkowski is a research leader in AI and artificial life. He is the founding director of Cross Labs, a research institute in Kyoto, Japan, focusing on the fundamental principles of intelligence in biological and synthetic systems. He is the president of the International Society for Artificial Life (ISAL). He is also an executive officer at leading AI company Cross Compass Ltd., and a lecturer in information sciences at the Graduate School of Arts and Sciences of the University of Tokyo. He has cofounded various ventures in science and technology on three continents, including YHouse Inc. (a nonprofit transdisciplinary research institute in New York, on the emergence of consciousness in the universe) and the Center for the Study of Apparent Selves (with a focus on AI ethics and Wisdom traditions). His research focuses on collective, open-ended, and empathic AI paradigms built from a mathematical understanding of intelligence in any substrate.

ILLUSTRATION CREDITS

The illustrations featured as frontispieces for the essays in this book are drawn from the 49th International Children's Exhibition of Fine Arts Lidice (ICEFA Lidice), held in 2021. This extraordinary annual exhibition is organized by the Lidice Memorial, an organization of the Ministry of Culture of the Czech Republic. The organization is dedicated to keeping alive the memory of the annihilation of the village of Lidice, razed to the ground by German Nazi troops on June 10, 1942. The annual exhibition was established to commemorate the child victims of Lidice, as well as other children whose lives have been affected or ended by warfare. The exhibition became international in 1973, and in recent years more than 25,000 children from over 60 countries have participated.

In 2021 the theme of the exhibition was robots and artificial intelligence. The catalog stated: "Since the time Karel Čapek wrote the drama *R.U.R. (Rossum's Universal Robots)*, advanced technologies have become a part of everyday life and the word 'robot' has spread out all over the world. Various machines and devices help people in many ways. However, they bring about both unprecedented opportunities and unprecedented pitfalls. All this is reflected in works by young artists presented in this year's exhibition."

We are grateful to ICEFA Lidice, and to the individual artists, for their permission to reproduce the works in this book. We encourage you to explore more at https://www.mdvv-lidice.cz/en.

Essay 1   **104**
Ariana Peťkovská, age 8, Častkovce, Slovak Republic

Essay 2   **116**
Andrea Gatialová, age 8, Brno, Czech Republic

Essay 3   **120**
Konstantin Mastikin, age 12, Birobidjan, Russia

Essay 4   **132**
Radim Špeta, age 10, Písek, Czech Republic

Essay 5   **146**
Kate Cirša, age 10, Limbaži, Latvia

NOTES

JITKA ČEJKOVÁ / INTRODUCTION

The editor would like to acknowledge the networking support of the COST Actions CA17120—Chemobrionics and CA21169—Information, Coding, and Biological Function: The Dynamics of Life (DYNALIFE), supported by COST (European Cooperation in Science and Technology).

1. H. Sayama, "ALife in a Larger Scientific Context," ALife Roadmap Workshop, ALIFE 2018, Tokyo, 2018.

2. J. Čejková, M. Novák, F. Štěpánek, and M. M. Hanczyc, "Dynamics of Chemotactic Droplets in Salt Concentration Gradients," *Langmuir* 30(40) (2014): 11937–11944.

3. J. Čejková, M. M. Hanczyc, and F. Štěpánek, "Multi-Armed Droplets as Shape-Changing Protocells," *Artificial Life* 24(1) (2018): 71–79.

4. J. Čejková, K. Schwarzenberger, K. Eckert, and S. Tanaka (2019), "Dancing Performance of Organic Droplets in Aqueous Surfactant Solutions," *Colloids and Surfaces A: Physicochemical and Engineering Aspects* 566 (2019): 141–147.

5. J. Čejková, T. Banno, F. Štěpánek, and M. M. Hanczyc (2017), "Droplets as Liquid Robots," *Artificial Life* 23(4) (2017): 528–549.

ŠTĚPÁN S. ŠIMEK / TRANSLATOR'S NOTE

1. At the September 2022 Tesla AI Day, Musk presented the current development of the humanoid robot Optimus, and stated, "This means a future of abundance. . . . A future where there is no poverty. Where . . . you can have whatever you want in terms of products and services." Faiz Siddiqui, "Elon Musk Debuts Tesla Robot, Optimus, Calling It a 'Fundamental Transformation,'" *Washington Post*, September 30, 2022, updated October 1, 2022, https://www.washingtonpost.com/technology/2022/09/30/elon-musk-tesla-bot/.

JULYAN CARTWRIGHT / ROBOTS AND THE PRECOCIOUS BIRTH OF SYNTHETIC BIOLOGY

1. Norma Contrada, "Golem and Robot: A Search for Connections," *Journal of the Fantastic in the Arts* 7(2/3) (1995): 244–254.

2. Karel Čapek, in the *London Evening Standard*, June 2, 1924.

3. Robert Kanigel, *The One Best Way: Frederick Winslow Taylor and the Enigma of Efficiency* (New York: Viking, 1997).

4. Frank Barkley Copley, *Frederick W. Taylor: Father of Scientific Management*, vol. 1 (New York: Harper and Brothers, 1923).

5. Frederick Winslow Taylor, *The Principles of Scientific Management* (New York: Harper & Brothers, 1911), 59.

6. Alan Myers, "Evgenii Zamiatin in Newcastle," *Slavonic and East European Review* 68(1) (1990): 91–99.

7. Leonardo Torres Quevedo, "Ensayos sobre automática. Su definición. Extensión teórica de sus aplicaciones," *Revista, Ciencias Exactas, Academia de ciencias* 12 (1913): 391–419.

8. Anonymous, "Torres and His Remarkable Automatic Devices," *Scientific American* 80(2079) (1915): 296–298.

9. Arturo Rosenblueth, Norbert Wiener, and Julian Bigelow, "Behavior, Purpose and Teleology," *Philosophy of Science* 10 (1943): 18–24.

10. Norbert Wiener, *Cybernetics: Or Control and Communication in the Animal and the Machine* (Cambridge, MA: MIT Press, 1948).

11. Isaac Asimov, *The Complete Robot* (New York: Doubleday, 1982).

12. H. G. Wells, "Mr. Wells Reviews a Current Film: He Takes Issue with This German Conception of What the City of One Hundred Years Hence Will Be Like," *New York Times*, April 17, 1927.

13. Norbert Wiener, *The Human Use of Human Beings* (London: Eyre & Spottiswoode, 1950).

14. Mark E. Rosheim, *Leonardo's Lost Robots* (Heidelberg: Springer, 2006); Mario Taddei, ed., *Leonardo da Vinci's Robots* (Milan: Leonardo3, 2007).

15. Joseph Seligmann Kohn, *Der jüdische Gil Blas* (Leipzig, 1834).

16. Norbert Wiener, *God & Golem, Inc.* (Cambridge, MA: MIT Press, 1964).

17. Contrada, "Golem and Robot."

18. Johann R. Glauber, *Furni novi philosophici* (1646), trans. Christopher Packe (London, 1689).

19. Isaac Newton, "Of Natures Obvious Laws & Pocesses in Vegetation" (c. 1675), Dibner Collection MS. 1031 B, The Dibner Library of the History of Science and Technology, Smithsonian Institution Libraries, https://pages.dlib.indiana.edu/concern/scanned_resources/d8336hb57s#?c=0&m=0&s=0&cv=3&xywh=28%2C1422%2C21 11%2C1666.

20. Moritz Traube, "Experimente zur Theorie der Zellenbildung und Endosmose," *Archiv für Anatomie, Physiologie und wissenschaftliche Medicin* (1867): 87–165.

21. Wilhelm Pfeffer, *Osmotische Untersuchungen* (1877); English translation: *Osmotic Investigations: Studies on Cell Mechanics*, trans. Gordon R. Kepner and Eduard J. Stadelmann (New York: Van Nostrand Reinhold, 1985).

22. Alfonso Luis Herrera, *Nociones de biología* (1903); facsimile ed., Universidad Autónoma de Puebla, Mexico, 1992.

23. Stéphane Leduc, *The Mechanism of Life* (London: Rebman, 1911).

24. Karl Marx to Pyotr Lavrov, June 18, 1875, in Karl Marx and Frederick Engels, *Collected Works*, vol. 45: *Letters, 1874–79* (London: Lawrence & Wishart, 2010).

25. Benjamin C. Gruenberg, "The Creation of 'Artificial Life,'" *Scientific American* 105(11) (1911): 231–237.

26. Elizabeth Bishop, *North and South* (Boston: Houghton Mifflin, 1946).

27. Thomas Mann, *Doktor Faustus* (Stockholm: Bermann-Fischer, 1947), 19.

28. R. D. Coatman, N. L. Thomas, and D. D. Double, "Studies of the Growth of 'Silicate Gardens' and Related Phenomena," *Journal of Materials Science* 15 (1980): 2017–2026.

29. Laura M. Barge, Silvana S. S. Cardoso, Julyan H. E. Cartwright, et al., "From Chemical Gardens to Chemobrionics," *Chemical Reviews* 115(16) (2015): 8652–8703.

30. Charles R. Darwin, *On the Origin of Species* (London: John Murray, 1859), 490.

31. Charles R. Darwin, letter to J. D. Hooker, February 1, 1871, in *The Correspondence of Charles Darwin*, ed. Frederick Burkhardt, James A. Secord, et al., vol. 19 (Cambridge: Cambridge University Press, 2012).

32. Michael J. Russell, Allan J. Hall, and Dugald Turner, "In Vitro Growth of Iron Sulphide Chimneys: Possible Culture Chambers for Origin-of-Life Experiments," *Terra Nova* 1(3) (1989): 238–241; Julyan H. E. Cartwright and Michael J. Russell, "The Origin of Life: The Submarine Alkaline Vent Theory at 30," *Interface Fous* 9(6) (2019): 2042–8898.

33. Yang Ding, Julyan H. E. Cartwright, and S. S. S. Cardoso, "Intrinsic Concentration Cycles and High Ion Fluxes in Self-Assembled Precipitate Membranes," *Interface Focus* 9(6) (2019): 2042–8898.

34. Leduc, *The Mechanism of Life*, 170.

35. John Desmond Bernal, *The Physical Basis of Life* (London: Routledge and Kegan Paul, 1951).

36. Simon Duval, Elbert Branscomb, Fabienne Trolard, et al., "On the Why's and How's of Clay Minerals' Importance in Life's Emergence," *Applied Clay Science* 195 (2020): 105737.

37. Silvana S. S. Cardoso, Julyan H. E. Cartwright, Jitka Čejková, et al., "Chemobrionics: From Self-Assembled Material Architectures to the Origin of Life," *Artificial Life* 26(3) (2020): 315–326.

38. Clyde A. Hutchison III, Ray-Yuan Chuang, Vladimir. N. Noskov, et al., "Design and Synthesis of a Minimal Bacterial Genome," *Science* 351(6280) (March 25, 2016).

39. Pasquale Stano, "Minimal Cellular Models for Origins-of-Life Studies and Bio-technology," in *The Biophysics of Cell Membranes*, ed. Richard M. Epand and Jean-Marie Ruysschaert (Heidelberg: Springer, 2017), 177–219.

CARLOS GERSHENSON / HUMANS AND MACHINES: DIFFERENCES AND SIMILARITIES

1. Marvin Minsky, *The Society of Mind* (New York: Simon and Schuster, 1986), 163.

INMAN HARVEY / *R.U.R.* AND THE ROBOT REVOLUTION: INTELLIGENCE AND LABOR, SOCIETY AND AUTONOMY

1. Adrienne Mayor, *Gods and Robots: Myths, Machines, and Ancient Dreams of Technology* (Princeton: Princeton University Press, 2018).

2. Eva Anagnostou-Laoutides and Alan Dorin, "The Silver Triton," *Nuncius* 33(1) (2018): 1–24.

3. David Silver, Thomas Hubert, Julian Schrittwieser, et al., "A General Reinforcement Learning Algorithm That Masters Chess, Shogi, and Go through Self-Play," *Science* 362(6419) (2018): 1140–1144, DOI: 10.1126/science.aar6404.

4. Inman Harvey, Ezequiel Di Paulo, Rachel Wood, Matt Quinn, and Elio Tuci, "Evolutionary Robotics: A New Science for Studying Cognition," *Artificial Life* 11(1–2) (2005): 79–98, DOI: 10.1162/1064546053278991.

5. Matthew D. Egbert, Jean Sébastien Gagnon, and Juan Pérez-Mercader, "From Chemical Soup to Computing Circuit: Transforming a Contiguous Chemical Medium into a Logic Gate Network by Modulating Its External Conditions," *Journal of the Royal Society Interface* 16(158) (2019), DOI: 10.1098/rsif.2019.0190.

6. Randall D. Beer and John C. Gallagher, "Evolving Dynamical Neural Networks for Adaptive Behavior," *Adaptive Behavior* 1(1) (1992): 91–122, DOI: 10.1177/10597 1239200100105.

7. Robert J. Shiller, *Narrative Economics: How Stories Go Viral and Drive Major Economic Events* (Princeton: Princeton University Press, 2019).

8. Harvey et al., "Evolutionary Robotics."

9. David Harris Smith and Frauke Zeller, "The Death and Lives of hitchBOT: The Design and Implementation of a Hitchhiking Robot," *Leonardo* 50(1) (2017): 77–78, DOI: 10.1162/LEON_a_01354.

10. Norman White, "Helpless Robot" installation, on compart: Center of Excellence Digital Art, accessed October 17, 2022, http://dada.compart-bremen.de/item/artwork/609.

11. Inman Harvey, "Evolving Robot Consciousness: The Easy Problems and the Rest," in *Consciousness Evolving*, ed. Hames H. Fetzer (Amsterdam: John Benjamins, 2002), 205–219.

12. Alan M. Turing, "Computing Machinery and Intelligence," *Mind* 59(236) (1950): 433–460, DOI: 10.1093/mind/LIX.236.433.

JULIE NOVÁKOVÁ / IT WASN'T WRONG TO DREAM: THE PARADISE OR HELL OF OUR JOBLESS FUTURE

1. Carl Benedict and Michael A. Osborne, "The Future of Employment: How Susceptible Are Jobs to Computerization?," *Technological Forecasting and Social Change* 114 (2017): 254–280, DOI: 10.1016/j.techfore.2016.08.019.

2. Derek Thompson, "A World without Work," *Atlantic*, July/August 2015 (accessed January 10, 2020), https://www.theatlantic.com/magazine/archive/2015/07/world-without-work/395294/.

3. Scott Santens, "The Expanse's Basic Support vs. Basic Income: How Not to Meet Humanity's Basic Needs in Our Automated Future," *Scott Santens*, September 22, 2017 (accessed January 10, 2020), http://www/scottsantens.com/the-expanse-basic-support-basic-income.

4. Research depository, *BIEN: Basic Income Earth Network* (accessed January 10, 2020),https://basicincome.org/research/research-depository/.

5. Studies: Single summary of projects and pilots, *The Basic Income Community on Reddit*, Reddit, June 26, 2016 (accessed January 10, 2020), https://www/reddit.com/r/BasicIncome/wiki/studies.

6. Scott Santens, "Universal Basic Income Will Accelerate Innovation by Reducing Our Fear of Failure," *Medium*, December 1, 2016 (accessed January 10, 2020), https://medium.com/basic-income/universal-basic-income-will-accelerate-innovation-by-reducing-our-fear-of-failure-b81ee65a254.

7. David K. Evans and Anna Popova, "Cash Transfers and Temptation Goods," *Economic Development and Cultural Change* 12(2) (2017): 20170, DOI: 10.1086/689575.

8. Karl Widerquist, "The Cost of Basic Income: Back-of-the-Envelope Calculations," *Basic Income Studies* 12(2) (2017): 20170, DOI: 10.1515/bis-2017–0016.

LANA SINAPAYEN / *R.U.R.*: A SHREWD PLUTOCRAT, A GENIUS ENGINEER, AND AN ANTI-SUE WALK INTO A BAR

1. Karel Čapek, *R. U. R.*, translated by Paul Selver and Nigel Playfair (New York: Dover Publications, 2001), Kindle Edition.

2. Jack Dorsey was the CEO of Twitter in October 2019 when this essay was written.

GEORGE MUSSER / ARTIFICIAL PANPSYCHISM

1. Karl Marx, *Capital*, trans. Ben Fowkes (New York: Penguin, 1990), 548.

2. David J. Chalmers, "Could a Large Language Model Be Conscious?," lecture, Conference on Neural Information Processing Systems, New Orleans, November 28, 2022.

3. William James, *The Principles of Psychology* (New York: Holt, 1890), 146–150.

4. Giulio Tononi, "Consciousness as Integrated Information: A Provisional Manifesto," *Biological Bulletin* 215, no. 3 (December 2008): 216–242, DOI: 10.2307/25470707.

5. Michael D. Kirchhoff and Tom Froese, "Where There Is Life There Is Mind: In Support of a Strong Life-Mind Continuity Thesis," *Entropy* 19, no. 4 (April 14, 2017): 169, DOI: 10.3390/e19040169.

6. Stanislas Dehaene, *Consciousness and the Brain: Deciphering How the Brain Codes Our Thoughts* (New York: Penguin, 2014), 100–109, 113–114, 244–249.

7. Rudy Rucker, "Panpsychism Proved," *Nature* 439, no. 7075 (January 26, 2006): 508, DOI: 10.1038/439508a

8. Rudy Rucker, "Everything Is Alive," *Progress of Theoretical Physics Supplement* 173 (2008): 363–370, DOI: 10.1143/PTPS.173.363.

9. Rudy Rucker, *Postsingular* (New York: Tor, 2007).

10. Rudy Rucker, *Hylozoic* (New York: Tor, 2009).

11. Rudy Rucker, "Writing Notes for *Hylozoic*," *Rudyrucker.com*, February 25, 2009, 81, https://www.rudyrucker.com/hylozoic/.

12. Erik Hoel, "AI Makes Animists of Us All," *The Intrinsic Perspective*, February 16, 2022, https://open.substack.com/pub/erikhoel/p/ai-makes-animists-of-us-all.

13. David Abram, *The Spell of the Sensuous* (New York: Pantheon, 1996).

JANA HORÁKOVÁ / THE ROBOT

1. Jana Horáková and Jozef Kelemen, "The Robot Story: Why Robots Were Born and How They Grew Up," in *The Mechanical Mind in History*, ed. Philip Husbands, Owen Holland, and Michael Wheeler (Cambridge, MA: MIT Press, 2008), 283–306.

2. Jana Horáková, *Robot jako Robot* (Prague: KLP, 2010).

3. Karel Čapek, "O slově robot," *Lidové noviny*, December 24, 1933, 12.

4. Robert Wechsler, "Karel Čapek in America," in *On Karel Čapek: A Michigan Slavic Symposium*, ed. Michael Makin and Jindřich Toman (Ann Arbor: UMI Research Press, 1992), 109–125.

5. Karel Čapek, "R.U.R.," *Prager Tagblatt*, September 23, 1935.

6. Josef Čapek, *Ledacos: fejetony [Umělý člověk]* (Prague: Dauphin, 1997), 196.

7. Umberto Eco, *The Limits of Interpretation* (Bloomington: Indiana University Press, 1991), 185.

8. Michael Holquist, "How to Play Utopia: Some Brief Notes on the Distinctivness of the Utopian Fiction," in *Game, Play and Literature*, ed. Jacques Ehrmann (Boston: Beacon Press, 1968), 106–203.

9. Greimas was influenced by Saussure's linguistic distinction between *langue* and *parole* and by Saussure's and Jakobson's concept of binary opposition.

10. Quoted in Terence Hawkes, "Algirdas Julien Greimas," in *Strukturalismus a sémiotika* (Brno: Host, 1999), 73–78.

11. Otokar Fischer, "Roboti a lidé (K premiéře kolektivního dramatu RUR od Karla Čapka)," *Jeviště* 2(3) (1921): 37–38.

12. Jiří Opelík, *Josef Čapek* (Prague: Melantrich, 1980), 28.

13. Daniela Hodrová, . . . *na okraji chaosu* . . . *Poetika literárního díla 20. Století* (Prague: Torst, 2001), 587.

ANTOINE PASQUALI / SCIENCE WITHOUT CONSCIENCE IS THE SOUL'S PERDITION

1. David Hilbert, *Natur und Mathematisches Erkennen*, ed. D. E. Rowe (Basel: Birkhäuser, 1992).

2. Julien Offray de la Mettrie, *Machine Man and Other Writings*, ed. A. Thomson (Cambridge: Cambridge University Press, 2003).

3. Carl Nägeli, "Zellenkerne, Zellenbindung und Zellenwachsthum bei den Pflanzen: Fortsetzung und Schluss," *Zeitschrift für wissenschaftliche Botanik* 3–4 (1846): 22–93.

4. Gregor Johann Mendel, "Versuche über Pflanzen-Hybriden," *Verhandlungen des naturforschenden Vereines in Brünn* 4 (1866): 3–47; reprinted in *Fundamenta Genetica*, ed. J. Kříženecký (Prague: Czech Academy of Sciences, 1966), 15–56.

5. Francis Crick and Christof Koch, "Towards a Neurobiological Theory of Consciousness," *Seminars in Neuroscience* 2 (1990): 263–275.

6. Antonio Damasio, *The Feeling of What Happens: Body and Emotion in the Making of Consciousness* (New York: Harcourt Brace, 1999).

7. Antonio Damasio, *Looking for Spinoza: Joy, Sorrow, and the Feeling Brain* (New York: Houghton Mifflin Harcourt, 2003).

HIROKI SAYAMA / KAREL ČAPEK: THE VISIONARY OF ARTIFICIAL INTELLIGENCE AND ARTIFICIAL LIFE

1. Christopher G. Langton, ed., *Artificial Life: Proceedings of an Interdisciplinary Workshop on the Synthesis and Simulation of Living Systems* (Boston: Addison-Wesley Longman, 1989); Christopher G. Langton, Charles Taylor, Doyne Farmer, and Steen Rasmussen, eds., *Artificial Life II: Proceedings of the 2nd Interdisciplinary Workshop on the Synthesis and Simulation of Living Systems* (Boston: Addison-Wesley Longman, 1991); Christopher G. Langton, ed., *Artificial Life III: Proceedings of the 3rd Interdisciplinary Workshop on the Synthesis and Simulation of Living Systems* (Reading, MA: Addison-Wesley Longman, 1993).

2. See http://alife.org/conferences.

3. Takashi Ikegami, "Chemical Robot: Self-Organizing Self-Moving Oil Droplet," *Journal of the Robotics Society of Japan* 28(4) (2010): 435–444, DOI: 10.7210/jrsj.28.435; Jitka Čejková, Taisuke Banno, Martin M. Hanczyc, and František Štěpánek, "Droplets as Liquid Robots," *Artificial Life* 23(4) (2017): 528–549, DOI: 10.1162/ARTL_a_00243.

4. Mark A. Bedau, John S. McCaskill, Norman H. Packard, et al., "Open Problems in Artificial Life," *Artificial Life* 6(4) (2006): 528–549, DOI: 10.7210/jrsj.28.435.

JOSH BONGARD / ROSSUM'S UNIVERSAL XENOBOTS

1. Sam Kriegman, Douglas Blackiston, Michael Levin, and Josh Bongard, "A Scalable Pipeline for Designing Reconfigurable Organisms," *Proceedings of the National Academy of Sciences* 117(4) (2020): 1853–1859, DOI: 10.1073/pnas.1910837117.

DOMINIQUE CHEN / THE LESSON OF AFFECTION FROM THE WEAK ROBOTS

1. Norbert Wiener, *The Human Use of Human Beings: Cybernetics and Society* (Boston: Houghton Mifflin, 1950), 181.

2. Francesco J. Varela, Evan Thompson, and Eleanor Rosch, *The Embodied Mind: Cognitive Science and Human Experience* (Cambridge, MA: MIT Press, 1991).

3. Dominique Chen, Young Ah Seong, Hiraku Ogura, Yuto Mitani, Naoto Sekiya, and Kiichi Moriya, "Nukabot: Design of Care for Human-Microbe Relationships," in Extended Abstracts of the 2021 CHI Conference on Human Factors in Computing Systems (CHI EA '21). Association for Computing Machinery, New York, Article 291, 1–7, https://doi-org.waseda.idm.oclc.org/10.1145/3411763.3451605.

4. Gregory Bateson, *Steps to an Ecology of Mind: Collected Essays in Anthropology, Psychiatry, Evolutions and Epistemology* (Chicago: University of Chicago Press, 2000).

TAKASHI IKEGAMI / GENERATIVE ETHICS IN ARTIFICIAL LIFE

1. Morris Berman, *The Reenchantment of the World* (Ithaca: Cornell University Press, 1981), 275.

GEOFF NITSCHKE AND GUSZ EIBEN / FROM *R.U.R.* TO ROBOT EVOLUTION

1. A. Mayor, *Gods and Robots: Myths, Machines, and Ancient Dreams of Technology* (Princeton: Princeton University Press, 2018).

2. J. von Neumann, *Theory of Self-Reproducing Automata* (Champaign: University of Illinois Press, 1966).

3. R. Freitas, *Kinematic Self-Replicating Machines* (Boca Raton: CRC Press, 2004).

4. L. Brodbeck, S. Hauser, and F. Iida, "Morphological Evolution of Physical Robots through Model-Free Phenotype Development," *PLoS ONE* 10(6) (2015):e0128444, DOI:10.1371/journal.pone.0128444; M. Hale, E. Buchanan, A. Winfield, et al., "The ARE Robot Fabricator: How to (Re)produce Robots that Can Evolve in the Real World," in *Proceedings of the 2019 Conference on Artificial Life* (Cambridge, MA: MIT Press, 2019), 95–102; M. Jelisavcic, M. De Carlo, E. Hupkes, et al., "Real-World Evolution of Robot Morphologies: A Proof of Concept," *Artificial Life* 23(2) (2017): 206–235; D. Howard, A. Eiben, D. Kennedy, J.-B. Mouret, P. Valencia, and D. Winkler, "Evolving Embodied Intelligence from Materials to Machines," *Nature Machine Intelligence* 1(12) (2019): 12–19.

5. D. Coldewey, "Robots Date, Mate, and Procreate 3D Printed Offspring in Robot Baby Project," *Techcrunch*, May 31, 2016, https://techcrunch.com/2016/05/31/robots-date-mate-and-procreate-3d-printed-offspring-in-robot-baby-project/; M. Simon,

"Robot 'Natural Selection' Recombines into Something Totally New," *Wired*, March 26, 2019, https://www.wired.com/story/how-we-reproduce-robots/; E. Hart, "Meet the Robots that Can Reproduce, Learn and Evolve All by Themselves," *New Scientist*, February 23, 2022, https://www.newscientist.com/article/mg25333751-700-meet-the -robots-that-can-reproduce-learn-and-evolve-all-by-themselves/.

6. H. Lipson and J. Pollack, "Automatic Design and Manufacture of Robotic Life Forms," *Nature* 406(1) (2000): 974–978.

7. R. Pfeifer and C. Scheier, *Understanding Intelligence* (Cambridge, MA: MIT Press, 1999). This book covers the study of artificial intelligence and its subfields such as machine learning in the context of embodied (physical robot) systems.

8. S. Doncieux, N. Bredeche, J.-B. Mouret, and A. Eiben, "Evolutionary Robotics: What, Why, and Where To," *Frontiers in Robotics and AI* 2(4) (2015), DOI:10.3389/ frobt.2015.00004.

9. F. Silva, L. Correia, and A. Christensen, "Evolutionary Online Behaviour Learning and Adaptation in Real Robots," *Royal Society* 4(160938) (2017).

10. J. Long, *Darwin's Devices: What Evolving Robots Can Teach Us about the History of Life and the Future of Technology* (New York: Basic Books, 2012); L. Brodbeck, S. Hauser, and F. Iida, "Morphological Evolution of Physical Robots through Model-Free Phenotype Development," *PLoS ONE* 10(6) (2017): e0128444, DOI:10.1371/ journal.pone.0128444; M. Jelisavcic, M. De Carlo, E. Hupkes, et al., "Real-World Evolution of Robot Morphologies: A Proof of Concept," *Artificial Life* 23(2) (2017): 206–235; M. Hale, E. Buchanan, A. Winfield, et al., "The ARE Robot Fabricator: How to (Re)produce Robots that Can Evolve in the Real World," in *Proceedings of the 2019 Conference on Artificial Life*, 95–102. In these cases, morphological mutation still requires some manual construction by engineers.

11. R. Pfeifer and J. Bongard, *How the Body Shapes the Way We Think* (Cambridge, MA: MIT Press, 2006).

12. Čapek's decision to make the robots indistinguishable from humans was likely not due to any prescience about future robotics, but rather to accommodate the practical limitations of stage plays in the early twentieth century and thus the necessity of having human actors assume roles of the robots.

13. A. Eiben, S. Kernbach, and E. Haasdijk, "Embodied Artificial Evolution: Artificial Evolutionary Systems in the 21st Century," *Evolutionary Intelligence* 5(1) (2012): 261–272.

14. A. Eiben, N. Bredeche, M. Hoogendoorn, et al., "The Triangle of Life: Evolving Robots in Real-Time and Real-Space," in *Proceedings of the 12th European Conference on the Synthesis and Simulation of Living Systems* (ECAL 2013) (Cambridge, MA: MIT Press, 2013), 1056–1063.

15. A. Eiben and J. Smith, "From Evolutionary Computation to the Evolution of Things," *Nature* 521(1) (2015): 476–482.

16. A. E. Eiben, "EvoSphere: The World of Robot Evolution," *Theory and Practice of Natural Computing* (TPNC 2015): 3–19, https://doi.org/10.1007/978-3-319-26841-5_1.

17. A. Eiben, E. Hart, J. Timmis, A. Tyrrell, and A. Winfield, "Towards Autonomous Robot Evolution," in *Software Engineering for Robotics*, ed. A. Cavalcanti, B. Dongol, R. Hierons, J. Timmis, and J. Woodcock (Cham: Springer, 2021), 29–51.

18. D. Howard, A. Eiben, D. Kennedy, J.-B. Mouret, P. Valencia, and D. Winkler, "Evolving Embodied Intelligence from Materials to Machines," *Nature Machine Intelligence* 1(12) (2019): 12–19.

19. J.-B. Mouret and K. Chatzilygeroudis, "20 Years of Reality Gap: A Few Thoughts about Simulators in Evolutionary Robotics," in *Proceedings of the 2017 Genetic and Evolutionary Computation Conference Companion* (GECCO '17) (New York: Association for Computing Machinery, 2017), 1121–1124.

20. J. M. Smith, "Byte-Sized Evolution," *Nature* 355(1) (1992): 772–773.

21. Eiben and Smith, "From Evolutionary Computation to the Evolution of Things."

22. M. Crichton, *Prey* (New York: HarperCollins, 2002).

23. T. Wallin, J. Pikul, and R. Shepherd, "3D Printing of Soft Robotic Systems," *Nature Reviews Materials* 3(1) (2018): 84–100.

24. G. Nitschke and D. Howard, "Autofac: The Perpetual Robot Machine," *IEEE Transactions on Artificial Intelligence* 3(1) (2022): 2–10.

25. J. Bellingham and K. Rajan, "Robotics in Remote and Hostile Environments," *Science* 318(5853) (2007): 1098–1102.

26. M. Sabatini and G. Palmerini, "Collective Control of Spacecraft Swarms for Space Exploration," *Celestial Mechanics and Dynamical Astronomy* 105(1) (2009): 229–244.

27. M. Couceiro, "An Overview of Swarm Robotics for Search and Rescue Applications," in *Handbook of Research on Design, Control, and Modeling of Swarm Robotics*, ed. Ying Tan (Hershey: IGI Global, 2016).

28. V. Jorge, R. Granada, R. Maidana, D. Jurak, G. Heck, A. Negreiros, D. dos Santos, L. Goncalves, and A. Amory, "A Survey on Unmanned Surface Vehicles for Disaster Robotics: Main Challenges and Directions," *Sensors* 19(3) (2019): 1–10.

29. B. Bayat, N. Crasta, A. Crespi, A. Pascoal, and A. Ijspeert, "Environmental Monitoring Using Autonomous Vehicles: A Survey of Recent Searching Techniques," *Current Opinion in Biotechnology* 45(1) (2017): 76–84.

30. J. Lewis, *Mining the Sky: Untold Riches from the Asteroids, Comets, and Planets* (New York: Perseus, 1997).

MIGUEL AGUILERA AND IÑIGO R. ARANDIA / ROBOTS AT THE EDGE OF CHAOS AND THE PHASE TRANSITIONS OF LIFE

1. Sara Imari Walker and Paul C. W. Davies, "The Algorithmic Origins of Life," *Journal of the Royal Society Interface* 10(79) (2013), DOI: 10.rsif.2012.0869; Thierry Mora and William Bialek, "Are Biological Systems Poised at Criticality?," *Journal of Statistical Physics* 144(2) (2011): 268–302, DOI: 10.1007/s10955-011-0229-4; Miguel A. Muñoz, "Colloquium: Criticality and Dynamical Scaling in Living Systems," *Review of Modern Physics* 8, 90(3) (2018), DOI: 10.1103/RevModPhys.90.031001.

2. Herbert A. Simon, "The Organization of Complex Systems," *Models of Discovery: and Other Topics in the Methods of Science* (Dordrecht: Springer, 1977), 245–261.

3. Christopher T. Kello and Guy C. Van Orden, "Soft-Assembly of Sensorimotor Function," *Nonlinear Dynamics Psychology and Life Sciences* 13(1) (2009): 57–78; Guy C. Van Orden, John G. Holden, and Michael R. Turvey, "Self-Organization of Cognitive Performance," *Journal of Experimental Psychology: General* 132(3) (2003): 331–350, DOI: 10.1037/0096-3445.132.3.331.

4. Mora and Bialek, "Are Biological Systems Poised at Criticality?"; Chris G. Langton, "Computation at the Edge of Chaos: Phase Transitions and Emergent Computation," *Physica D: Nonlinear Phenomena* 42(1–3) (1990): 12–37, DOI: 10.1016/0167-2789(90)90064-V.

5. Kello and Van Orden, "Soft-Assembly of Sensorimotor Function."

6. Giulio Tononi, Melanie Boly, Marcello Massimini, and Christof Koch, "Integrated Information Theory: From Consciousness to Its Physical Substrate," *Nature Reviews Neuroscience* 17(7) (2016): 450–461, DOI: 10.1038/nrn.2016.44.

7. Leo P. Kadanoff, "More Is the Same: Phase Transitions and Mean Field Theories," *Journal of Statistical Physics* 137(5–6) (2009): 777–797, DOI: 10.1007/s10955-009-9814-1.

8. Ezequiel A. Di Paolo, "Enactive Becoming," *Phenomenology and the Cognitive Sciences* (2020), DOI: 10.1007/s11097-019-09654-1.

9. Gilbert Simondon, *Individuation in Light of Notions of Form and Information*, trans. Taylor Adkins (Minneapolis: University of Minnesota Press, 2020).

10. Di Paolo, "Enactive Becoming," 792.

11. Simondon, *Individuation in Light of Notions of Form and Information*, 7.

12. Di Paolo, "Enactive Becoming."

OLAF WITKOWSKI / ROBOTIC LIFE BEYOND EARTH

1. Nikolai S. Kardashev, "Transmission of Information by Extraterrestrial Civilizations," *Soviet Astronomy* 8(2) (1964): 217–221.

2. Piet Hut, Walter Alvarez, William P. Elder, et al., "Comet Showers as a Cause of Mass Extinctions," *Nature* 329(6135) (1987): 118–126, DOI: 10.1038/329118a0.

3. Robert Marcus, H. Jay Melosh, and Gareth Collins, "Earth Impact Effects Program," *Impact: Earth!* [online] (London: Imperial College London, 2010) [cit. 2013-02-04].

4. Stephen Hawking, "Last Letters on the Future of Planet Earth," *Sunday Times*, October 14, 2018.

5. Olaf Witkowski, "Evolution of Coordination and Communication in Groups of Embodied Agents" (Ph.D. diss., University of Tokyo, 2015).

6. Stephen Hawking, *Brief Answers to the Big Questions* (New York: Random House, 2018), https://xn--webducation-dbb.com/wp-content/uploads/2019/01/Stephen-Hawking-Brief-Answers-to-the-Big-Questions-Random-House-Publishing-Group-2018.pdf.

INDEX